Acclaim

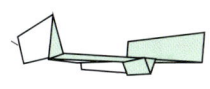

"*On Folded Wings* is a delightful presentation of Michael Weinstein's new origami for the paper airplane enthusiast. One of America's most dedicated designers, Weinstein has effectively interspersed well-chosen biographical accounts of aviation's most remarkable pioneers. These augment his many lessons about the mechanics and the physics of flight. Michael's origami designs are well conceived, fly well, and look great — the three most important aspects on our list! The clear, colorful illustrations are easy on the eyes and easier to follow. There is a generous section on origami symbols and techniques, and all of the models make use of clear landmarks to ensure reproducible results."

— *Michael G. LaFosse & Richard L. Alexander,
Origamido Studio, authors,* Planes for Brains.

"This book is a gem, highly recommended as an educational tool and as plain fun and games."

— *Tony Miksak, wordsonbooks.blogspot.com*

"*On Folded Wings* is one of the few paper airplane books designed by pilots. Michael Weinstein provides a great understanding of the history of flight and its evolution into modern-day aviation. He teaches you how to fold simple airplanes at first, and advance to more challenging ones as you learn. At the same time, he educates you on how real airplanes fly. Michael has folded some of the most intriguing and exciting models, and his beautiful designs aren't just for looks, but for aerodynamic high performance. This is a great book for anyone who loves folding and flying origami paper airplanes."

— *Lance Ty Yang, author of* Exotic Paper Airplanes *and* Exquisite Interceptors

"Michael has been doing great work with paper airplanes for decades. In this book, he pulls together the worlds of flight and pure origami as few authors can. In every way, he strives for the extraordinary: from the seldom-told tales of early flight, all the way through basic rocketry, there are ideas and planes enough for many happy hours. This is a book made to feed a paper-folding addiction. The only danger is that your favorite table will look like an aircraft carrier before you know it. Michael even seems to have thought of that, having designed a number of clever stands with which you can park your planes safely on bookshelves. Michael Weinstein belongs in the same class as James Sakoda and Stephen Weiss, American masters of origamic flight. *On Folded Wings* belongs on my favorites shelf."

— *John Collins, author of* The Gliding Flight, *and designer of
the 2012 World Record Indoor Paper Airplane for Distance*

Middleton Public Library
7425 Hubbard Avenue
Middleton, WI 53562

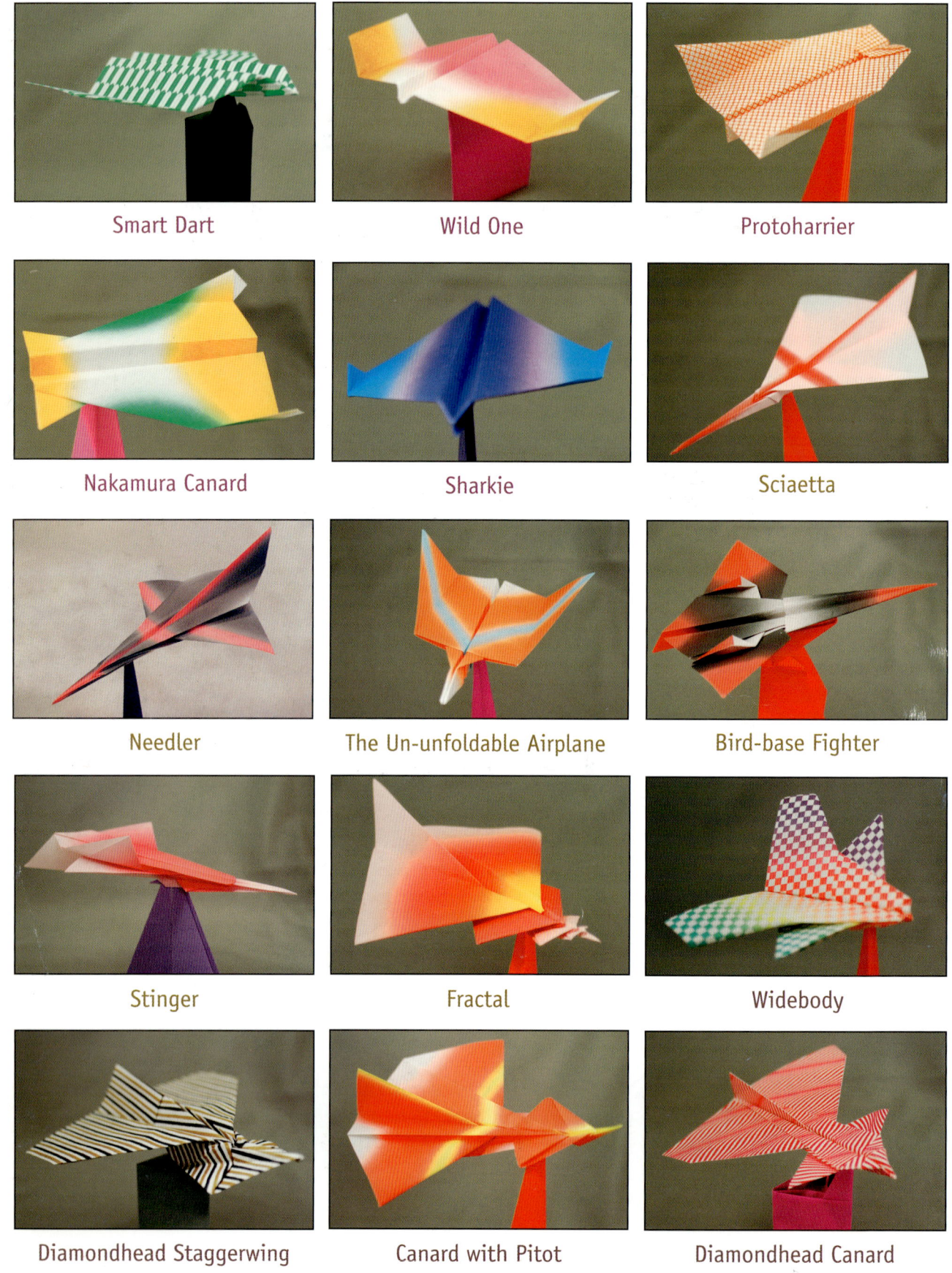

ON FOLDED WINGS

Paper Airplanes for All Ages

MICHAEL WEINSTEIN

Photographs by
John Martin Patrick Francis Aloysius Scully

Aircraft Illustrations by Mike Dietz

Fort Bragg, California

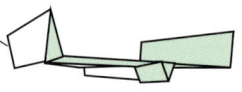

On Folded Wings
Paper Airplanes for All Ages

Copyright ©2013 by Michael Weinstein

All rights reserved. This book may not be reproduced in whole or in part, by any means, electronic or mechanical, without prior written permission from the publisher. For information, or to order additional copies of this book, please contact:

Cypress House
155 Cypress Street
Fort Bragg, CA 95437
(800) 773-7782
www.cypresshouse.com

Photographs by John Martin Patrick Francis Aloysius Scully
Aircraft Illustrations by Mike Dietz
Acrobat designed by Luigi Leonardi
Fractal image (Mandelbrot set) on p. 53 reproduced by permission of David Dewey: www.ddewey.net
Cover and text design by Michael Brechner/ Cypress House

Library of Congress Cataloging-in-Publication Data

Weinstein, Michael, 1962-
  On folded wings : paper airplanes for all ages / by Michael Weinstein.
     p. cm.
  ISBN 978-1-879384-79-8 (pbk. : alk. paper)
 1. Paper airplanes. 2. Aeronautics--History. I. Title.

TL778.W449 2010
745.592--dc22   2010014052

Printed in Hong Kong
2 4 6 8 9 7 5 3 1

Contents

INTRODUCTION • 1
Why Do Airplanes Fly? • 1
Why do paper airplanes fly the way they do? • 4
Bonus Airplane Number 1 • 4
Bonus Airplane Number 2 • 6

SYMBOLS • 17

GETTING STARTED • 21
Smart Dart • 22
Wild One • 24
Protoharrier • 26
Nakamura Canard • 28
Sharkie • 30
Sciaetta • 32

A FEW GOOD DARTS • 36
Needler • 37
The Un-unfold-able Airplane • 40
Bird-base Fighter • 43
Stinger • 47
Going Fast • 51
Fractal • 53

CANARDS • 56
Widebody • 57
Diamondhead Staggerwing • 60
Canard with Pitot • 63
Diamondhead Canard • 69
Twin Tail • 73

THE DROP ZONE • 76
Drop Zone • 77
Bird-base Glider • 79
Tail Dragger • 81
Gremlin • 85

BRANCUSI ANIMALS • 88
Enormously Abstract Heron • 89
Somewhat Less Abstract Canada Goose • 91
Not Entirely Abstract Kingfisher • 93
Owl in Space • 95

LOOPS, TUBES, AND ASSORTED MAYHEM • 98
Boxoid • 99
Triangulon • 101
Twin Star • 104
Nacelle Jet • 105

STANDS • 110
Stand Long • 111
Stand Tall • 113
Stand Open • 115
Stand Sharp • 116
Pylon Stand • 118
Hand Stand • 120

Introduction

Why Do Airplanes Fly?

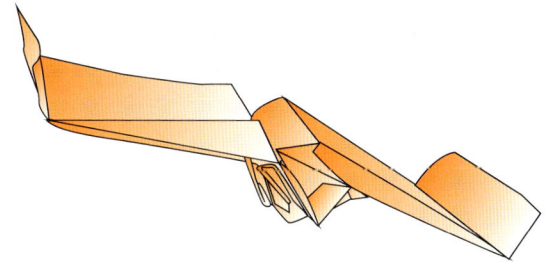

The simplest answer would be because we throw them. This, however, is not quite true. If you simply let go of one from atop a tall building, it will fly just fine. Indeed, Japanese astronauts plan to release small paper airplanes in space, to see if they can come down to planet Earth without burning, as meteors and the odd spaceship tend to do.

What throwing and dropping have in common is *forward momentum*. We give paper airplanes *thrust* by turning our own biological energy (from muscular contractions) into velocity when we throw them. Releasing them from a height is a bit more complicated, as we take the potential energy of the airplane being at an altitude (we had to use some sort of energy to get it up there) and turn it into kinetic energy as the airplane drops under the force of gravity. Either way the aircraft gains forward momentum.

To explain why this is a great thing we have to go back to one of the greatest scientists ever, Sir Isaac Newton. He was a truly amazing man, discovering differential calculus and the laws of physics at an age when most people can barely balance their checkbooks. If aliens have ever visited Earth, Newton was almost certainly one of them, as he was spectacularly intelligent.

Newton determined that a body in motion will remain so unless acted on by another force. But if that were the case our airplane would keep going in a straight line forever. We've already discussed one force that acts on airplanes, both paper and real: gravity. Also discovered by Newton, gravity pulls our airplane down as its momentum carries it skyward. A third force, drag, acts on our aircraft, and needs a little more Newtonian physics to explain.

Newton showed that for every action there is an equal and opposite reaction. As the airplane travels through the air, it hits lots and lots of air molecules, each collision slowing the airplane down a bit. Each air molecule only has a little force, but there are a lot of gas molecules in the air. If you're not convinced, try standing, walking, chewing gum, or doing anything else in a tornado. Put all these together and you have a force to be reckoned with, called drag by us aeronautical types. Everything that moves through the air creates drag; for us it's barely noticeable, but for paper airplanes it's considerably stronger.

**Sir Isaac Newton,
not just another pretty face**

Introduction

We have the beginnings of an explanation for why paper airplanes fly when we throw them. But we've also seen that they can fly when dropped from a height. From a tall building, drop a paper airplane along with a chainsaw, a banjo, and an accordion. Which one hits the ground first? Who cares?

Most things will fall straight down when dropped from a height. On the other hand, a properly made paper airplane will gain momentum and glide for a distance when dropped. To understand this phenomenon, we must think about Newton, as well as another long-deceased scientist named Bernoulli.

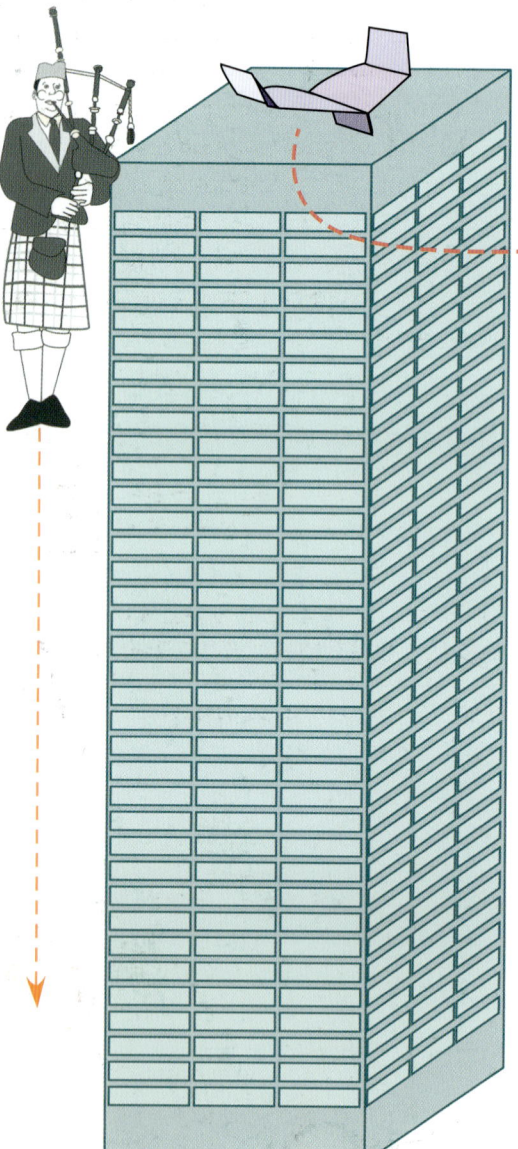

As we've noted, for every action there is an equal and opposite reaction. Think about a plane moving rapidly through the air. We've already seen that collisions with the molecules of the atmosphere reduce its momentum. But these collisions can also help keep the airplane flying.

In the figure below, an airplane is shown being hit by an air molecule (actually, it's being hit by a hydrogen atom—there are a few of them in the atmosphere). Notice that our brave molecule hits the wing and bounces off at an angle. The airplane must react, and the force exerted is shown by a solid line. Note that the force has both upward and backward components. The upward component goes by the friendly name of Lift, while the backward component is our old nemesis Drag. Again, this is easily tested; put your hand out the window of a moving vehicle (preferably not a commuter jet) and angle it upward. You'll notice a strong upward force. Lots of air molecules together exert a strong force.

That's one source of lift for airplanes, and is based on straightforward Newtonian physics. There is, however, another source of aerodynamic lift that's not so straightforward. To understand it, we really need to look at the shape of an airplane's wing. All airplanes have wings that are curved on top to at least a small extent. This curve, called camber, affects the travel of air molecules over it. To demonstrate this we show two air molecules racing to get across the surface of the wing.

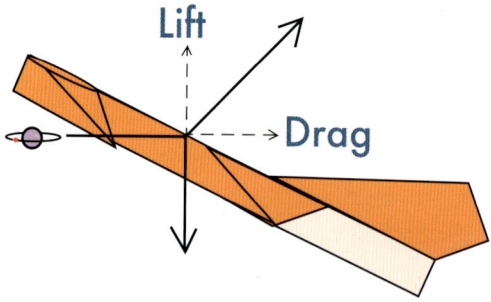

When an air molecule hits a wing (in this case of the Nakamura canard) it bounces off (solid arrow). The wing has an equal and opposite reaction (solid arrow) that can be divided into vertical and horizontal components.

Notice that the upper molecule has a longer way to go than its neighbor to the south has. Our deceased benefactor Bernoulli discovered that when a fluid speeds up, it exerts less pressure. Not convinced? Try holding your thumb over a garden hose. As the fluid emerges it both speeds up and loses pressure.

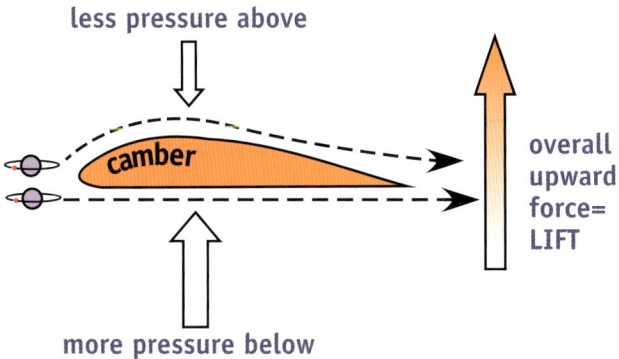

When air flows pasts a wing, the air on top must speed up to catch the air on the bottom. This causes a pressure differential that forces the wing upwards.

This was discovered for liquids, but it applies perfectly well to gasses, which are really weak-kneed wannabe liquids anyway. The air molecules on top go faster than the ones below, thus exerting less pressure. Less pressure above and more pressure below makes a net force upward, our pal Lift. To sum up, we get lift from two places: the incline of the wing, and the aerodynamic lift from the wing's camber. The other forces acting on an airplane are thrust (from throwing for a paper airplane, and the engines of real airplanes), gravity, and drag from collisions with atmospheric gasses.

But paper airplanes, like insects, have very, very thin wings. How is it that they generate lift? The answer comes from the work of Osborne Reynolds, a British engineer who studied fluid dynamics in the late 19th century. His work showed that air is indeed quite viscous for small objects like paper airplanes, and less so for larger ones, like us. To a paper airplane, air has more the consistency of water, and so the paper airplane can make enough lift to keep itself aloft with very thin wings. Think about how much more energy it takes to swim than walk; more viscous fluids also exert more drag. Greater camber produces

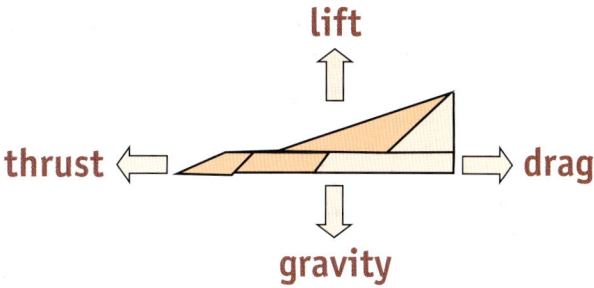

more drag—another reason why a paper airplane's wings must be thin.

Notice that something special happens when the wing is pointing quite high. The air molecules do not stick to the wing, and can pass along it. If there's not enough air on the wing, it will not have sufficient lift to keep the aircraft flying. Such a situation is referred to as an aerodynamic stall. When airplanes stall, it's not because the engine has quit (quite rare with most aircraft) but due to a lack of airflow over the wing. Some paper airplanes, such as Bonus Airplane Number 2 (coming up), stall out in flight due to a loss of momentum. They expend their upward momentum against atmospheric drag, then fall back down, gaining momentum and lift as they go.

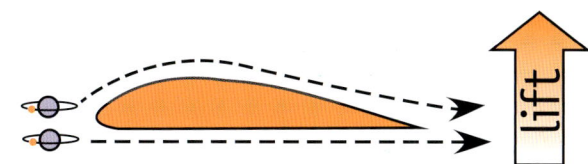

When a wing hits the airstream in level flight, lift is created through the previously described pressure differential. However, if the wing is at a steep enough incline, the air fails to adhere to its surface and it enters an aerodynamic stall.

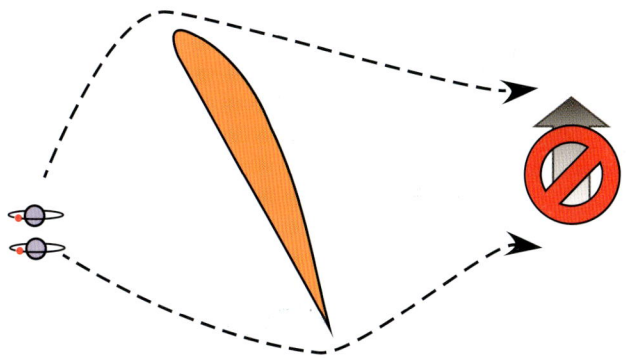

Introduction

Why do paper airplanes fly the way they do?

The best way to answer this and related questions is to build a couple of airplanes and see how they work.

Bonus Airplane Number 1

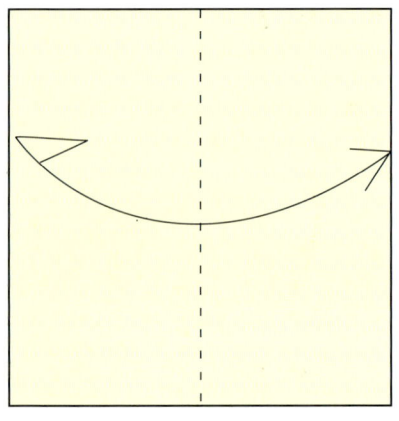

1. Begin by creasing a square in half lengthwise.

Why a square? I began using squares for airplanes because they presented an interesting challenge, being radially symmetrical. Cephalopods (squids, octopuses, and nautiluses) are radially symmetrical, and glide gracefully through the oceans of the world. Problem is, they flop about on land. Flopping about is generally frowned upon for aircraft. So how do we use a radially symmetrical square to make an aircraft? In this example, we'll pretend it's a normal rectangle, and treat it as such. We ignore many of the wonderful properties of a square, but that's all right, we'll use those later. Hey, it's only a bonus airplane! By creasing the square in half, we make it bilaterally symmetrical, like most aircraft. There have been a few airplanes that weren't bilaterally symmetrical, but these were decidedly silly. Although we humans are outwardly symmetrical, in point of fact we aren't. Our hearts point to the left, our spleens are on the right, and our guts are asymmetric (also gross and slimy). Failure to establish this left/right axis of symmetry during embryonic development can cause a disease called *situs inversus viscerum*, which is bad.

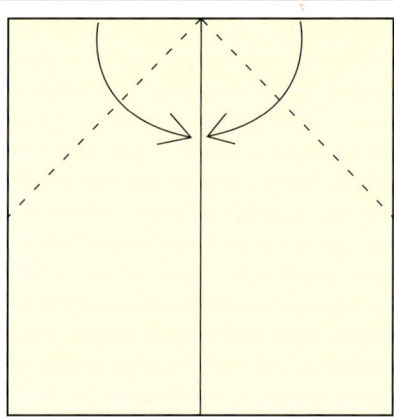

2. Fold the corners so that their upper raw edges lie on the centerline. In addition to giving you a wonderful sense of deja vu, we're narrowing the front end of the aircraft. Lots of airplanes are narrow at the front, as it allows them to slip more easily through the air.

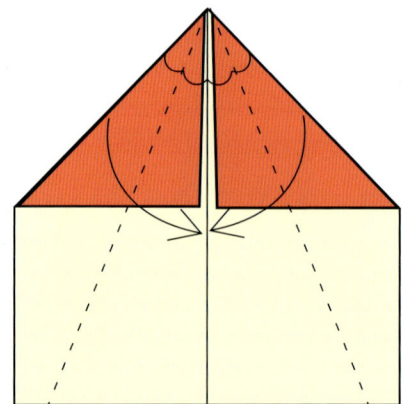

3. Valley fold again so that the folded edge up front lines up with the centerline. Again, we're making our airplane quite pointy. That's all right, lots of airplanes are pointy; it's sort of an airplane type thing.

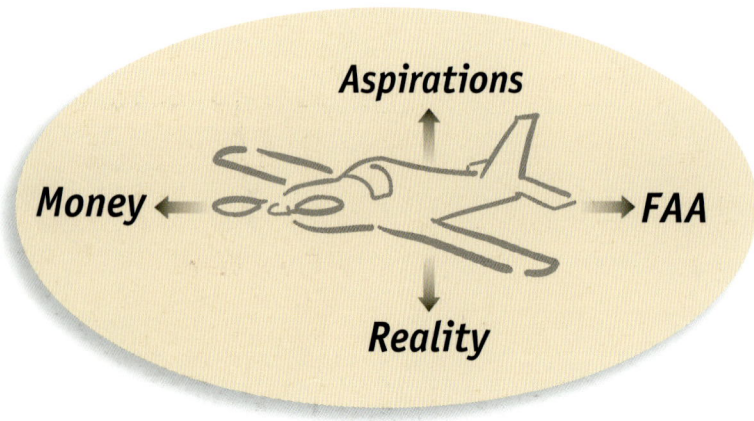

Bonus Airplane Number 1

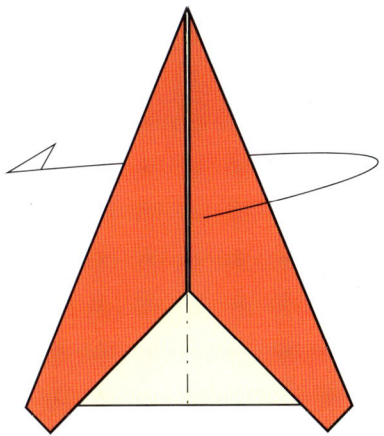

4. Mountain fold in half. Let's hear it for bilateral symmetry!

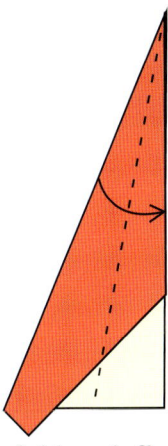

5. Now fold each flap in half to make wings. The middle of the paper will form the airplane's fuselage.

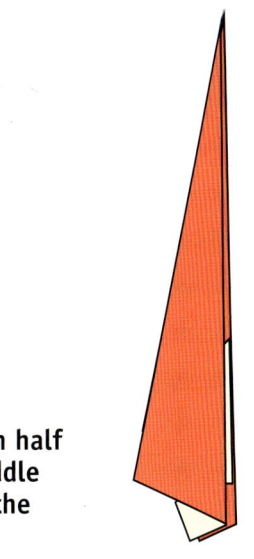

6. Now you're done. Notice that you get a front, side, and top view. These should allow you to figure out how it all relates. In this airplane the wings are perpendicular to the fuselage. Others may be somewhat different.

This airplane is probably one you're already familiar with, but we'll use it to show some of what makes airplane fly the way they do. I will take credit for its invention, however, as I designed it at a tender age. Indeed I was infinitely surprised that no one thought much of my novel invention. Alas, different inventors often make similar discoveries simultaneously. Now that you have your first airplane (no matter how original or not) give it a throw and check it out. Then go ahead and make Bonus Airplane Number 2. We'll compare their traits and behavior to learn more about what makes airplanes fly.

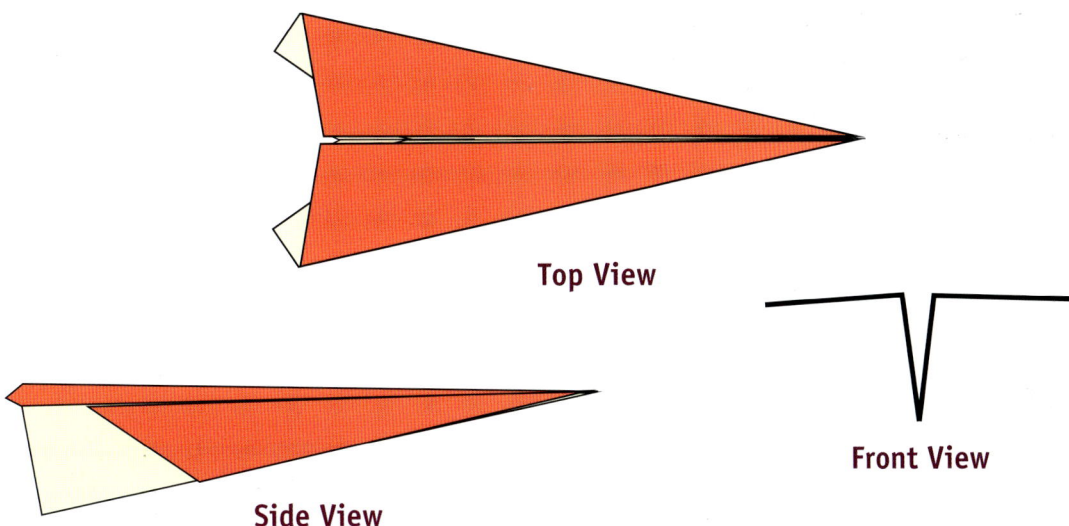

Top View

Side View

Front View

Introduction

Bonus Airplane Number 2

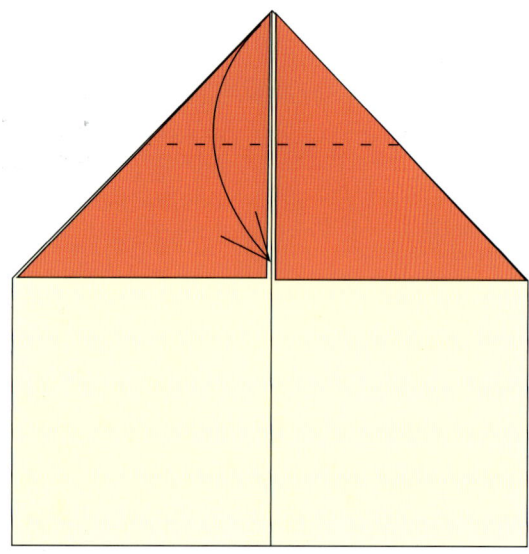

1. Begin with step 3 of Bonus Airplane Number 1. Valley fold the top point so that it touches the bottom of the colored section on the midline.

By now you've noticed that each fold has a definite relationship to either the paper or another fold. You'll find that true for most aircraft in this collection, as it allows you to fold each aircraft identically, and makes it easy to predict their flight characteristics. Real aircraft are created much the same way. Aircraft made by factories have type certificates to which they must adhere precisely, and for that reason are referred to as certificated. Many private pilots construct their own aircraft, based on plans or on kits manufactured by factories. Some airplanes don't conform to any type certificate; these are called experimental, a somewhat misleading moniker, as many so-called experimental aircraft are based on designs proven by time. Indeed, up to a third of the general aviation fleet is experimental. The *Spirit of St. Louis* was experimental; the *Hindenburg* certificated.

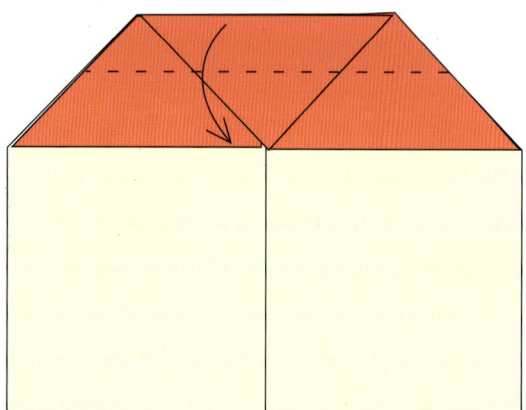

2. Valley fold the top so that it lies on the edge of the colored region.

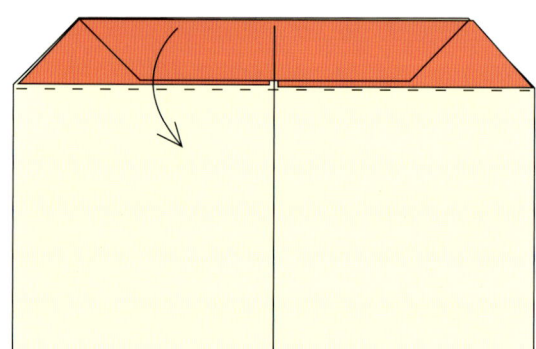

3. Valley fold the rest of the upper colored portion down.

An airplane might disappoint any pilot, but it'll never surprise a good one. — Len Morgan

Bonus Airplane Number 2

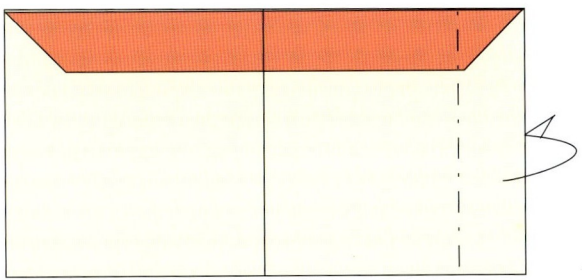

4. Mountain fold the end of the airplane at the point shown.

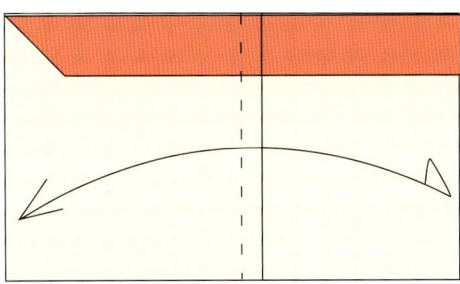

5. Valley fold so that the folded end of the airplane meets the other edge, and unfold.

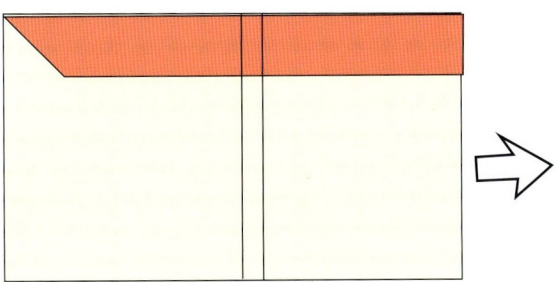

6. Unfold the edge.

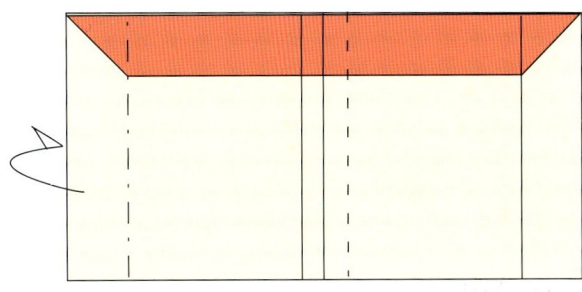

7. Repeat steps 4–6 on the other side.

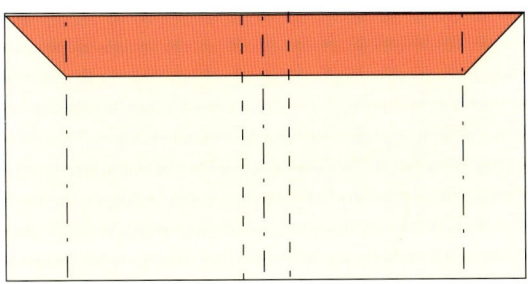

8. Mountain and valley fold along the creases; position the wings perpendicular to the fuselage, and the stabilizers on the ends perpendicular to the wings.

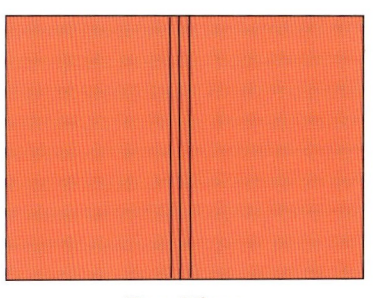

Top View

Side View

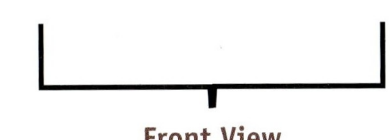

Front View

Eagles may soar, but weasels never get sucked into jet air intakes.
— *John Benfield*

Introduction

So now you have a pair of airplanes that look quite different. Give each a throw and see what happens! Obviously, they fly quite differently, which makes sense since they appear dissimilar. Why do these differences cause them to fly differently?

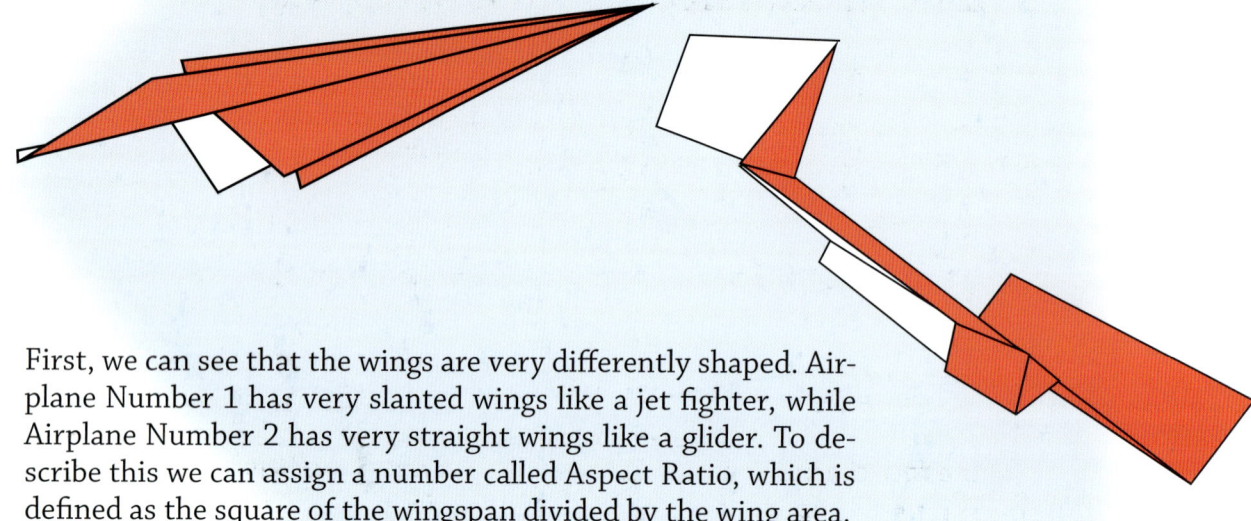

First, we can see that the wings are very differently shaped. Airplane Number 1 has very slanted wings like a jet fighter, while Airplane Number 2 has very straight wings like a glider. To describe this we can assign a number called Aspect Ratio, which is defined as the square of the wingspan divided by the wing area.

Really, this is a way of describing the relationship between a wing's length and width. Fast airplanes tend to have very low-aspect wings. These generate little lift but also make little drag, allowing the airplanes to move at higher velocities. High-performance sailplanes have very high-aspect wings that get the most out of the air. You're usually not in a hurry to go anywhere in a glider, but you do want to stay aloft for a long time.

If we're to understand aircraft at all, there's another aspect of airplane performance that we must dwell on: balance. A well-designed airplane, whether paper, cloth, wood, metal, composite, or graphite, must balance its weight with its lifting forces.

Aerodynamics

High-performance sailplanes can remain airborne with no motive power for hours at a time. In the American southwest they are routinely flown to altitudes so high that the pilots must use supplemental oxygen because the air is too thin.

Aerodynamics

A concept we must think about when discussing balance is Center of Gravity (CoG). This is an imaginary point found in all objects, at which the object is in balance. Airplanes, chairs, cephalopods, you and I—all have a center of gravity. To illustrate, let's think about a seesaw, which is really just a balance with a fulcrum in the center. If the weight on either end is the same, the seesaw should balance.

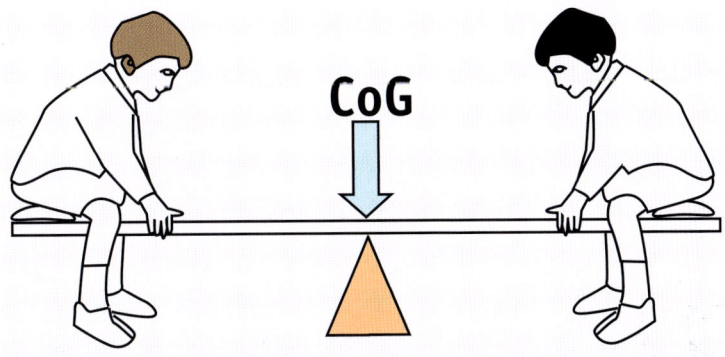

The fact that the seesaw balances means that its center of gravity is in the center. To illustrate this further, let's think about a seesaw with one large child and one small emergency-backup child.

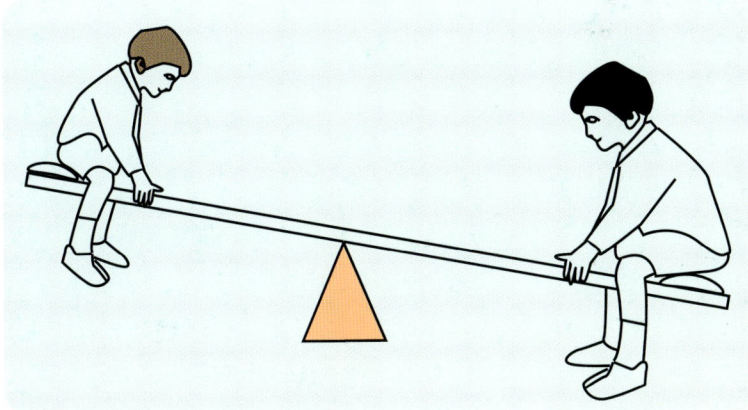

Notice that the seesaw is no longer balanced. (This always happened to me growing up, as I was a scrawny little kid on the way to growing into a scrawny little adult.) What has really happened with the seesaw is the center of gravity has moved. We can easily demonstrate this by moving the fulcrum.

Move the fulcrum over to the plus-sized child and the seesaw balances, because we have shifted its center of gravity. We changed the balance in both bonus airplanes by piling up layers of paper in front. Bonus Airplane Number 2 has more of this than Number 1, yet demonstrates more lift. Why?

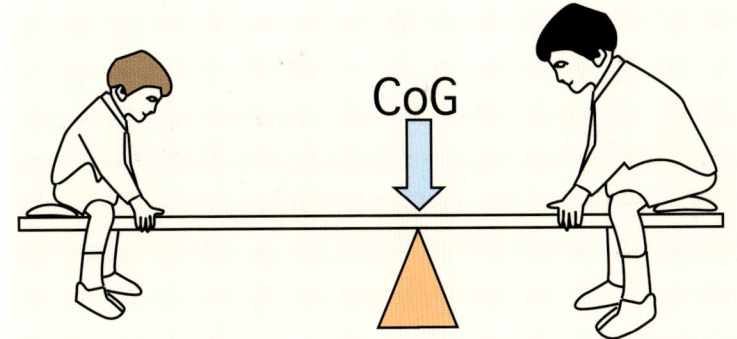

9

Introduction

Another concept we must master to complete our engineering analysis is the center of lifting forces. Just as every object has a center of gravity, every object has an imaginary point where all of the aerodynamic lifting forces acting on it are in balance. We don't worry about that too much unless we take up skydiving, but for airplanes it is vitally important.

Let's examine a normal airliner, and let's further confine ourselves to thinking about the airfoils (the wings and the elevators in the aft). This is somewhat simplistic, since most surfaces on an airplane exert some lift, but it will suffice. The wings produce the lion's share of the lift, shown by large arrows, but the elevators produce some as well, shown by smaller arrows. Therefore, the Center of Lift (CoL) must be somewhat aft of the wings. With four large, heavy jet engines forward of the wings, the center of gravity is likely to be forward of the wings or at their leading edge.

When we look at these forces relative to the airplane, we can see that the center of gravity is forward of the center of lift. This is of extreme importance for airplanes. If the center of gravity is aft of the center of lift, the airplane will have a tendency to nose up. This can result in a stall, which can lead to a precipitous loss of altitude and collision with the ground. For most airplanes, this is very, very bad.

The weight and balance of real airplanes can be affected by how their contents are loaded. This was demonstrated spectacularly when the singer and actress Aaliyah perished along with her band on August 25, 2001. The pilot of their chartered Cessna 402B inadvertently loaded the aircraft with too much weight in the back. Upon takeoff, the pilot was unable to lower the nose; the airplane stalled, crashed, and caught fire, as airplanes are wont to do. Unfortunately, the occupants, despite wishes to the contrary, were similarly flammable.

Pilots must always make certain that their aircraft are balanced. For example, I can't load too much weight in the back of my Piper Cherokee, or I'll suffer a fate similar to that of Aaliyah, her band, and their pilot. Similarly, we can't put two plus-sized people in front without putting a heavy weight in back, or the aircraft will be unable to raise its nose.

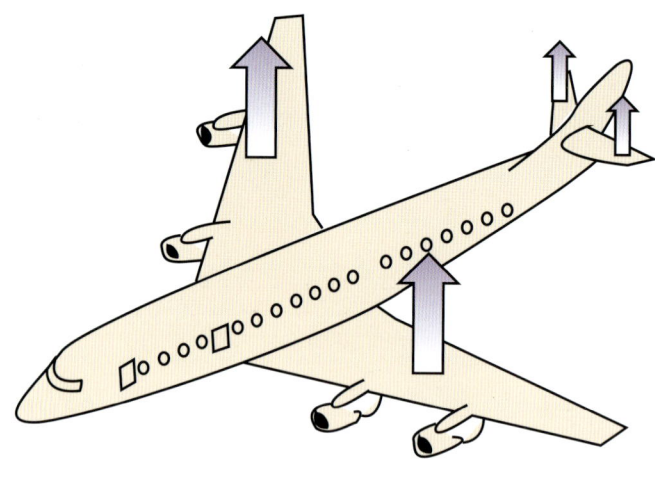
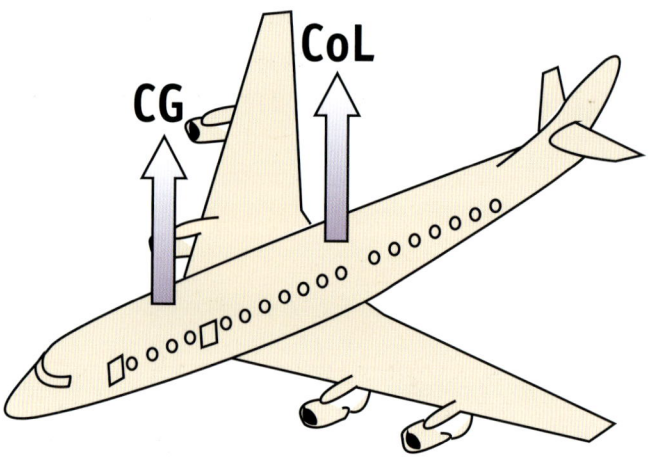

Airplanes make lift from their airfoils, pushing the Center of Lift (CoL) aft. However, their heavy jet engines move the Center of Gravity (CG) forward. The result is the CG is forward of the CoL.

Aerodynamics

We can see how these concepts apply to our paper airplanes. Number 1 has less paper in front than Number 2, thus its center of gravity will be aft of that in airplane number 2, which has more paper in front. On the other hand, Airplane Number 1 also has less wing surface forward, so its center of lift will be aft of Airplane Number 2 as well.

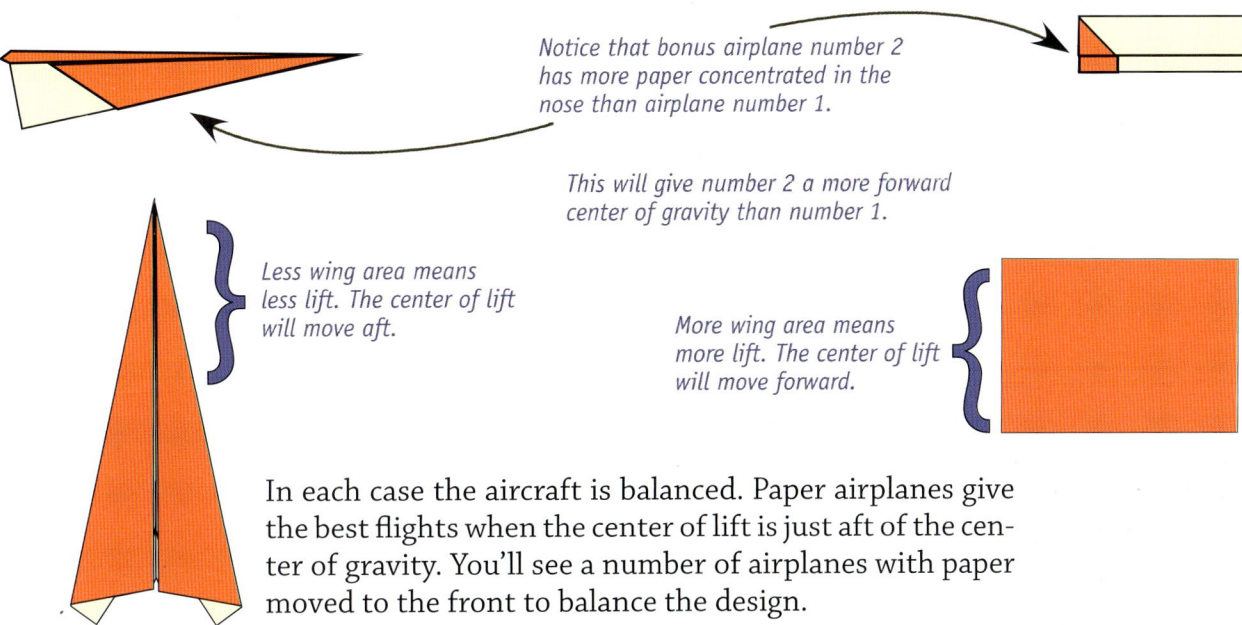

Notice that bonus airplane number 2 has more paper concentrated in the nose than airplane number 1.

This will give number 2 a more forward center of gravity than number 1.

Less wing area means less lift. The center of lift will move aft.

More wing area means more lift. The center of lift will move forward.

In each case the aircraft is balanced. Paper airplanes give the best flights when the center of lift is just aft of the center of gravity. You'll see a number of airplanes with paper moved to the front to balance the design.

How Can We Make Airplanes Fly the Way We Want?

All airplanes, whether made of paper, wood, cloth, metal, or fiberglass move through three axes, sort of like the X, Y, and Z axes for aircraft. These are pitch, roll, and yaw. Pitch is when the nose of the aircraft moves up or down, which is how pilots make aircraft ascend and descend. It works well; we haven't yet left a plane in the air. Roll is when the wings of the airplane go up or down, causing the whole thing to tilt. Despite tilting the aircraft, it's actually how pilots turn the aircraft left and right. Yaw is when the nose of the airplane moves back and forth, turning the aircraft left and right—except real airplanes use roll to turn left and right. Confused? Read on!

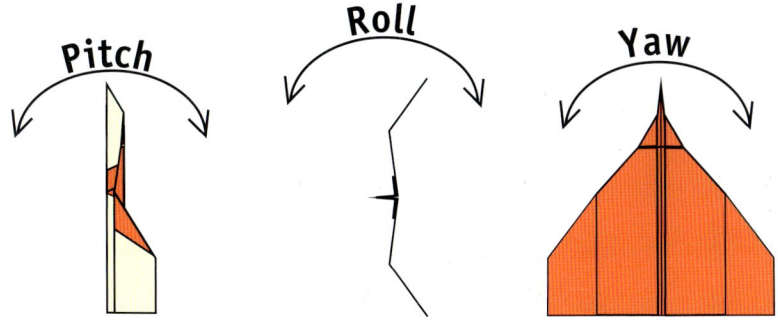

11

Introduction

Pilots obviously want airplanes to go where they desire. To achieve this, airplanes have control surfaces that direct them through the three axes of movement. Every airplane has some sort of elevator, usually a small wing in the back that controls pitch movements. All airplanes have roll control, often in the form of ailerons. These change the camber of the wings, allowing them to rise or fall. Most airplanes have a vertical stabilizer and rudder to control movements in the yaw axis. Flying wings do not, and are consequently unstable in yaw. The B2 bomber, which is a flying wing, requires sophisticated computer control to keep it stable.

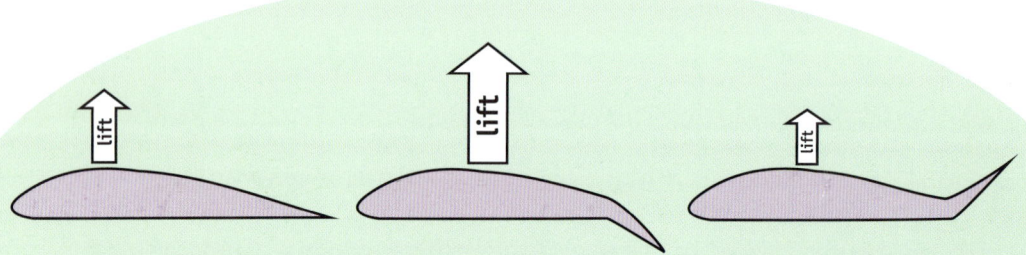

Ailerons can increase lift. Notice that when the aileron is down (middle) the camber of the wing increases (takes an air molecule longer to work its way around). When the aileron comes up, the lift is spoiled (right). Rudders and elevators work similarly, as do flaps, which are deployed to slow airplanes down prior to landing.

Ailerons function by changing the camber of the wings. If the aileron is deflected downward, the camber is increased, as is the lift. If the aileron is deflected upward, the lift is spoiled by the decrease in camber. When turning, one aileron goes up while the other goes down (they're interconnected). This causes one wing to ascend and its opposite member to descend.

Aircraft use rudders very differently from boats. Boats use rudders for turning; that is, changing direction in the yaw axis. But boats also change in the roll axis as they turn (the two are interconnected), indeed most things do. Motorcycle riders undergo huge changes in the roll axis when they turn. Riders lean their bikes into a turn, and racers do this so much they actually drag their knees on the ground. Many have special metal pucks on their knees so they can make sparks. Cars actually lean in the wrong direction, as they are on a platform. That's one of the main reasons a motorcycle can corner faster than a car.

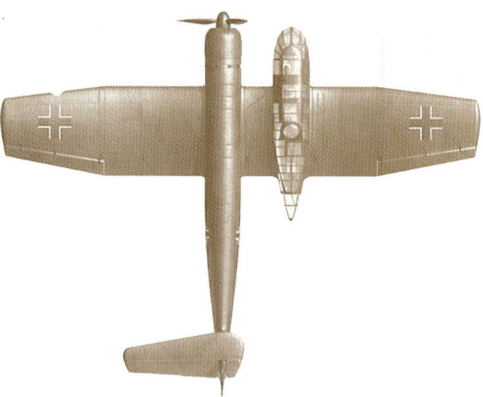

There have been asymmetric airplanes built, such as the Blohm and Voss 141. Designed by Dr. Richard Vogt in 1937, it proved too radical for the German Air Ministry and never saw production.

Airplanes do it a bit backwards. They begin a turn in the roll axis, banking the aircraft into the turn. The aircraft then turns in the yaw axis in consequence. The rudder doesn't turn the airplane; its only function in a real airplane is to coordinate turns. Airplanes can fly sideways quite effectively; such a maneuver is called a slip, and can be used to descend for landing.

The wings of a banked airplane actually make less lift than one flying straight and level. Therefore, left to its own devices, an airplane will descend as it goes into a turn. All three of the axes of movement are tied to each other. The airplane can be banked using rudder only, for example. Thus pilots have to learn how to control all three axes at once while flying.

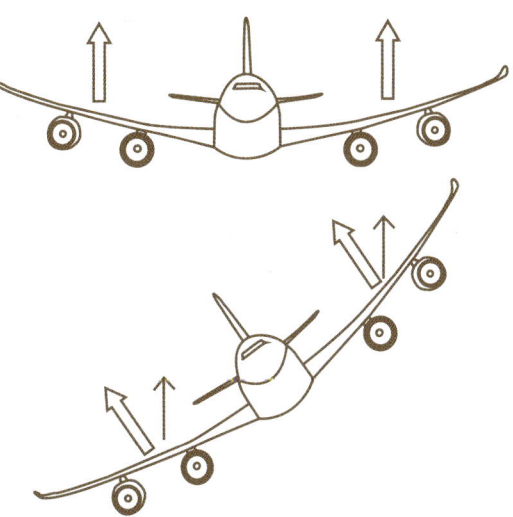

When an aircraft is flying wings level, the lift (thick arrows) is roughly perpendicular to the wings. When an airplane banks, the lift is still generated perpendicular to the wings, lessening its vertical component (thin arrows).

How Can We Make Paper Airplanes Fly the Way We Want?

Paper airplanes can have the same control surfaces as their fossil fuel-burning brethren. All the aircraft herein have horizontal and vertical surfaces that can act as ailerons, elevators, and rudders. Let's look at how to control the flight of Bonus Airplane Number 2.

By now you've seen that Bonus Airplane Number 2 doesn't fly so well; it rises, then stalls; rises, then stalls, and so on until it lands. I used it not because it flies well, but because it can be improved.

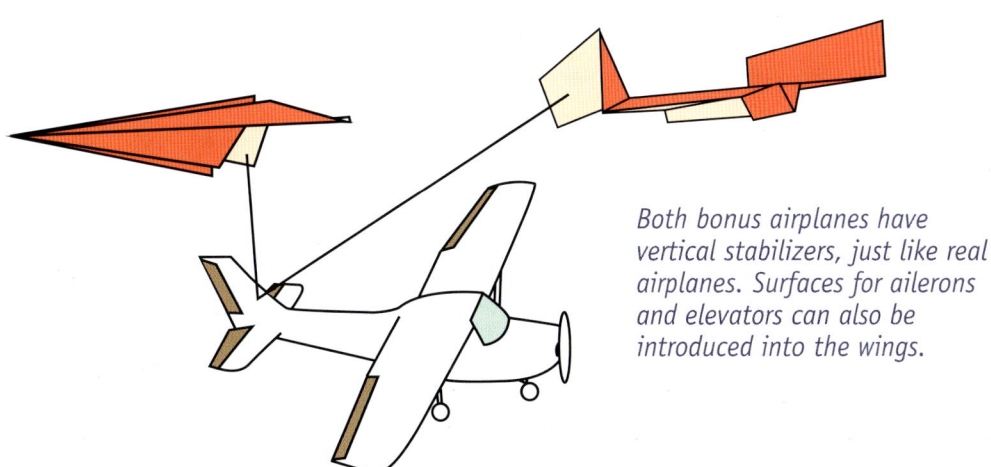

Both bonus airplanes have vertical stabilizers, just like real airplanes. Surfaces for ailerons and elevators can also be introduced into the wings.

Introduction

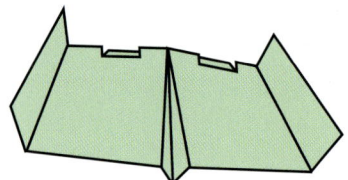

We should start by leveling off its flight a little bit. Remember: upward elevators can spoil the lift of a wing. But wait, we have no *empennage*, French for tail feathers, and the word for the rear wing and tail. We can still give Number 2 downward elevator by putting small tabs at the back of the wings. Despite being on the main wings, these will raise the trailing edge and level off the flight to some degree. The more elevator (wider, deeper, etc.) the more level the flight. Notice that it won't fly as far, because elevators introduce our old enemy, drag. But better a short satisfactory flight than a long unsatisfactory one.

Elevators can be added to the back of airplane #2 to level off the flight path.

You shouldn't have to do this on any of the other airplanes here; most have about the right amount of lift to fly nicely. However, you might use the opposite trick, that is giving the wing some upward elevator by bending the trailing edge downward, to get a bit more lift from some of the airplanes. Most won't need it, but depending on the paper and your folding technique you might find it helpful.

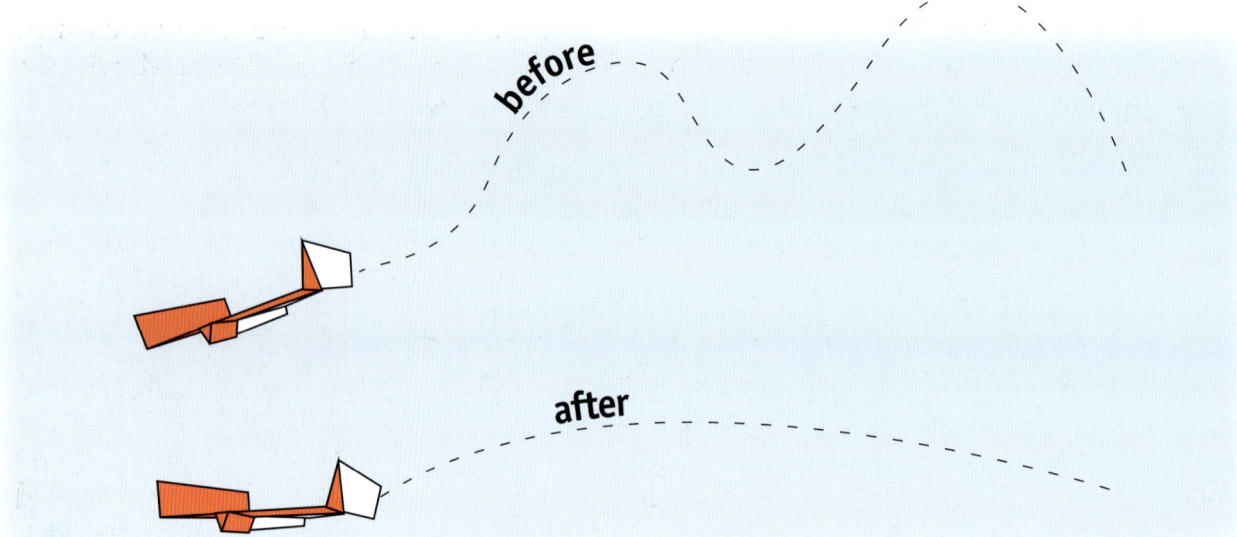

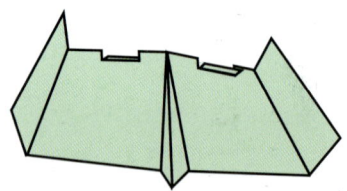

Ailerons can be added to the back of airplane #2 to cause it to circle.

You might also find it useful to be able to turn an airplane in flight. You may have limited space, or have religious convictions that can't countenance straight flight. Whatever the case, this is really easy to do with most airplanes from *On Folded Wings*. Remember, normal airplanes use ailerons to roll, which turn the airplane. Ailerons can be added to the main wings of Number 2 just like the elevator. The only difference is that instead of turning both down, one goes up and the other down (the up elevator should be in the direction you want the airplane to turn).

14

All right, you have the airplane turning. The Collier Trophy is yours. But what if you want it to turn more? Eeek, trophy in doubt! No worry. You can add more aileron, but too many folds are unsightly, and you like your airplane just the way it is. But wait—real airplanes have rudders in their vertical stabilizers. Your airplane has vertical stabilizers. Can you put rudders at their ends? Of course, and the rudders will tighten up a turn. Using a combination of aileron, rudder, and elevator you can make many of these airplanes do pretty much whatever you like.

One last word about making your paper airplanes fly the way you want. Most folks want an airplane with a reasonably stable flight envelope. There are exceptions. Fighter jets are actually very unstable, as are many acrobatic airplanes. Odds are you want your airplanes to be very stable in flight.

We've talked about achieving stability in the pitch axis (all that stuff about center of gravity). Now we'll talk about stability in the roll axis. One way to get it is to use dihedral. The idea is to angle the wings slightly upward. As the airplane leans over, the downward wing develops more lifting force than the upward one. When that happens, the downward wing will rise and the upward wing will fall. The airplane returns to level flight on its own accord.

Notice that you give up a bit of lift when you use dihedral on the wings, meaning your airplane won't fly as far. Many things in aircraft design are tradeoffs; you don't fly as far, but you also don't go into a spin and die. Fighter jets often are designed with anhedral, or a downward tilt to the wings. This is especially good for aircraft with extremely swept-back wings.

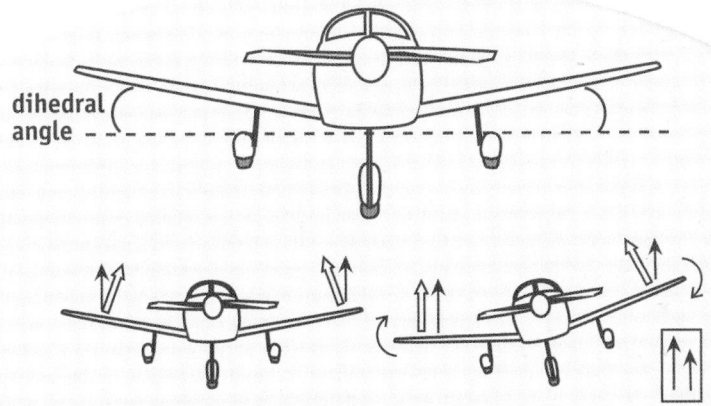

My Piper Cherokee is shown from the front. Notice the dihedral angle. What does this do? Remember, the lift (hollow arrows) is generated perpendicular to the wings. The upward component of the lift (thin arrows) is similar on both sides in level flight. In a roll the upward component of lift is substantially reduced on the upward wing compared to the lower wing (the two components are compared in the inset box). The increased lift below and decreased above causes a rolling motion (curved arrows) of the dihedral wing that returns the wing to straight and level flight. No such force acts on a straight wing, which will continue its roll if not acted upon. Thus aircraft with dihedral are more stable in the roll axis than ones without dihedral.

Introduction

What Do We Make Our Airplanes Out Of?

Probably sounds like a really stupid question. Actually, there are no stupid questions, only stupid politicians. Obviously, one wants to make a paper airplane out of paper. That covers a lot more ground than you might at first glance realize.

Most of the papers we use in our daily routine are quite suitable for folding these airplanes — copy, graph, and notebook paper will all work just fine. Newspaper will not; it's far too light and insufficiently crisp. Construction and art papers aren't entirely suitable either, as they're too heavy both for complex folds and long flights.

I'll make a special push for Japanese origami paper, called kami (the Japanese word for paper). It's light and crisp, and colored only on one side, just like my diagrams. I designed all the square airplanes in kami, and it is in kami that they work the best. There was a time when I wouldn't have done this, as origami paper used to be very difficult to get. In the age of the Internet, however, this is no longer true. Many bookstores, most hobby stores, and virtually all art stores carry kami, and there are numerous Internet vendors.

The best thing about kami is that there's an array of colors, patterns, and styles to choose from. Your airplanes can transcend their utilitarian origins to become diminutive works of art; indeed that's why you've been supplied with a selection of airplane stands as well.

Foiled!

Foil papers—laminates of thin metal foil and paper—are available to folders. Foil papers have a number of advantages over standard papers, not the least of which is that they hold their shape rather than springing back. That allows the folder to sculpt foil models in a realistic fashion.

The downside of foils is that once done, whatever is folded out of them looks like a robot whatever. Most things in nature, like cats, elephants, cephalopods, guitar players, and other organic beings aren't made of metal, and look somewhat unnatural when modeled in metallic paper.

Airplanes, on the other hand, are normally made of metal, and look quite natural made from foil. Unfortunately, foil has two distinct drawbacks when made into airplanes. It's heavier than kami and thus will increase wing loading, the amount of weight carried by a unit of wing area. More wing loading means shorter flight. The other is that foil has memory. It will "remember" mistakes and folds, and it's very difficult to get large areas, like wings, free of any defect. So foil will work for your airplanes, but it will have its disadvantages.

Airplane Designers in History

At the front of each section, I show some of the people who have designed aircraft. I've tried to show those that you may not have heard of. Most people recognize names like Wright and Lindbergh, fewer some of the people described in *On Folded Wings*. They've all contributed huge accomplishments, and all deserve to be remembered.

No bird soars too high, if he soars with his own wings. — William Blake

Symbols

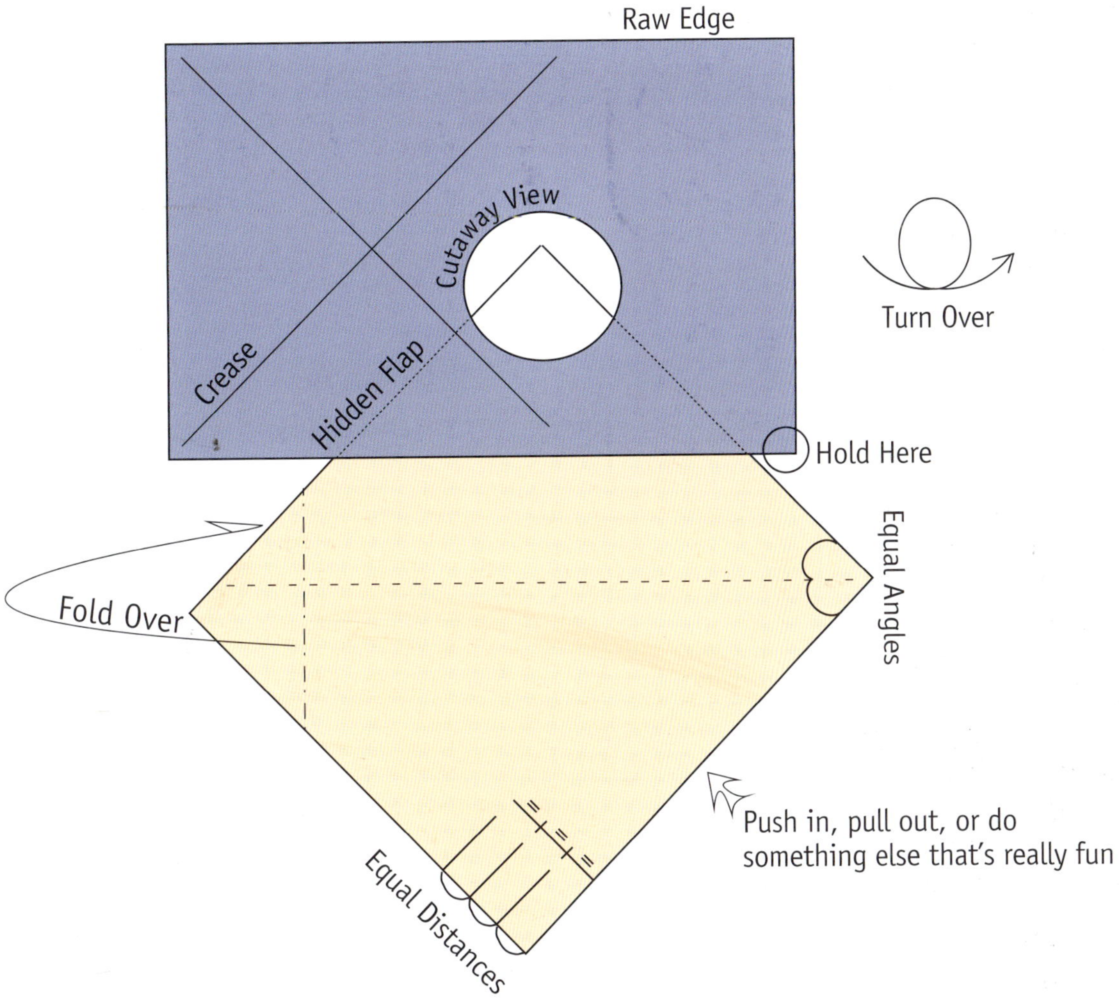

Raw Edge

Crease

Cutaway View

Hidden Flap

Turn Over

Hold Here

Fold Over

Equal Angles

Equal Distances

Push in, pull out, or do something else that's really fun

Valley Fold

Fold such that the crease points away from you (and the flap toward you).

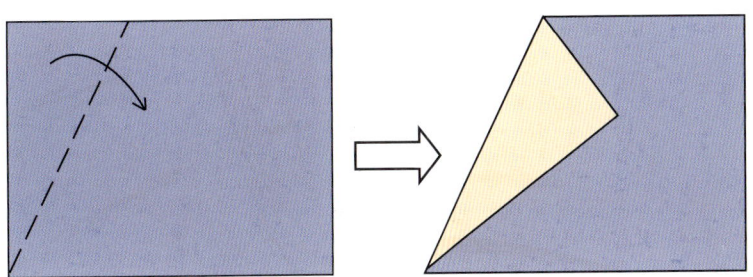

Introduction

Mountain Fold

Fold so that the crease points toward you (and the flap away from you).

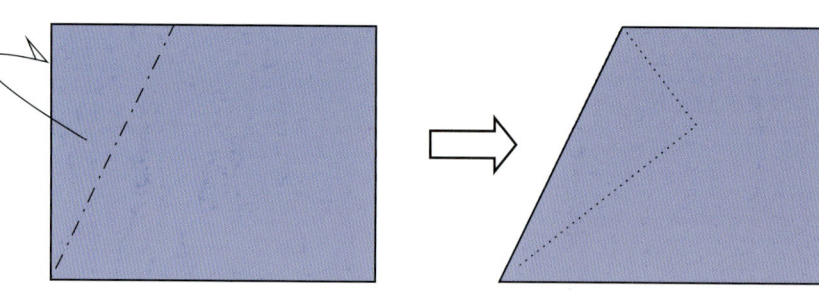

Inside Reverse Fold

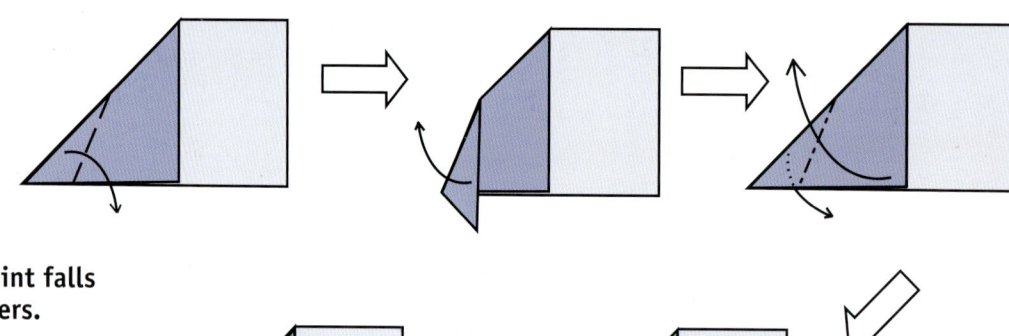

Fold so that a point falls between layers.

Outside Reverse Fold

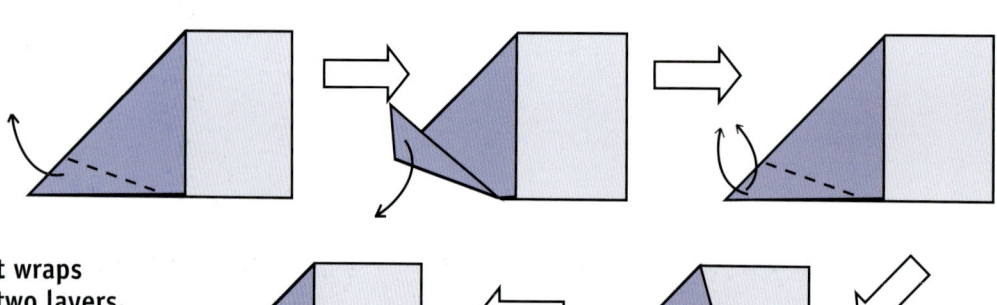

Fold so that a point wraps around the outside of two layers.

Rabbit Ear

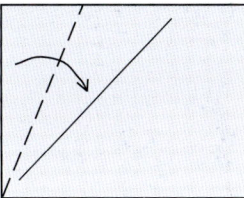
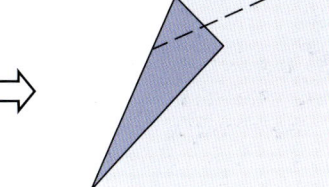

Pull out a hidden layer and fold it flat.

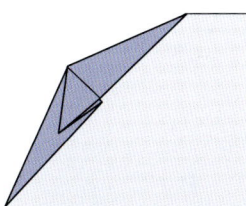
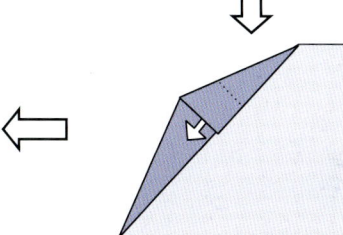

Squash Fold

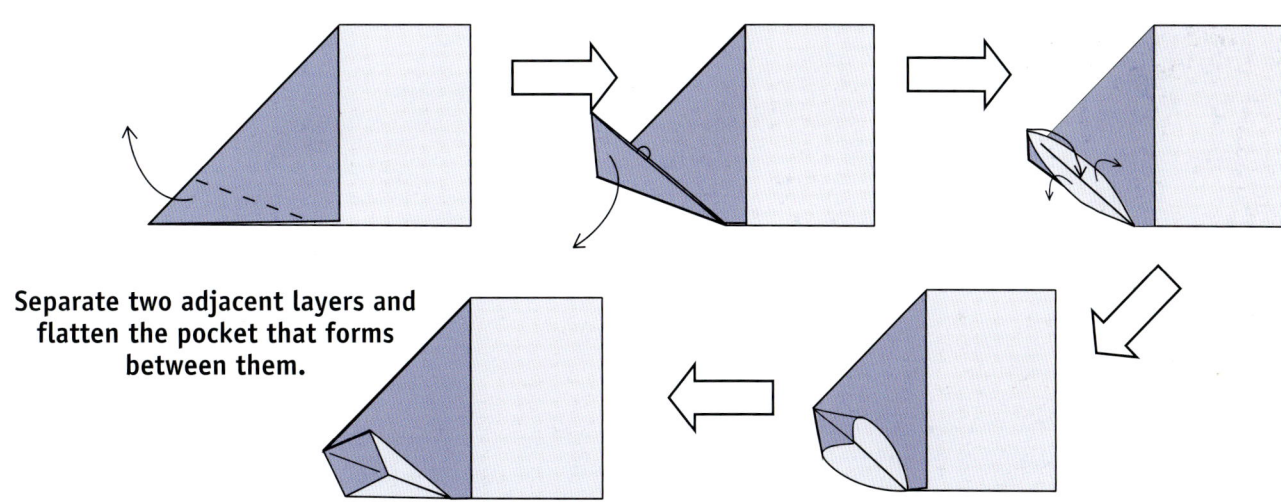

Separate two adjacent layers and flatten the pocket that forms between them.

You haven't seen a tree until you've seen its shadow from the sky. — Amelia Earhart

Introduction

Petal Fold

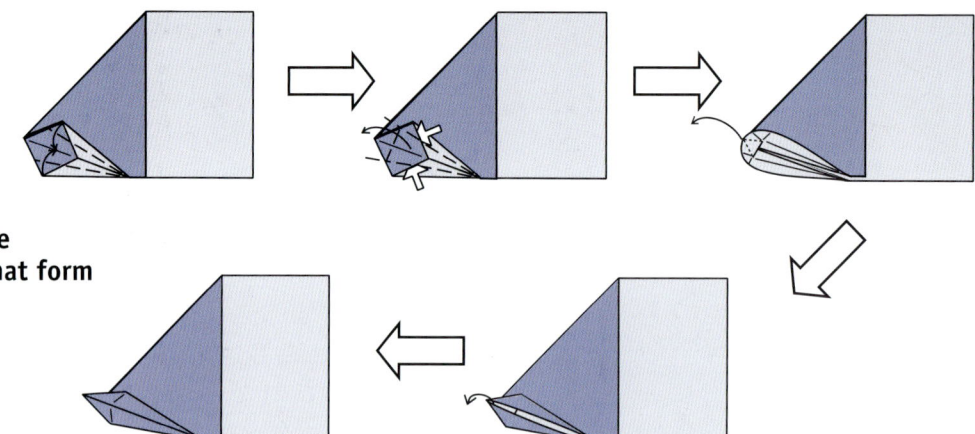

Separate two layers while flattening the pockets that form on either side.

Sink Fold

A multilayered point gets pushed to the inside. The point is partially unfolded and mountain folded at the desired position.

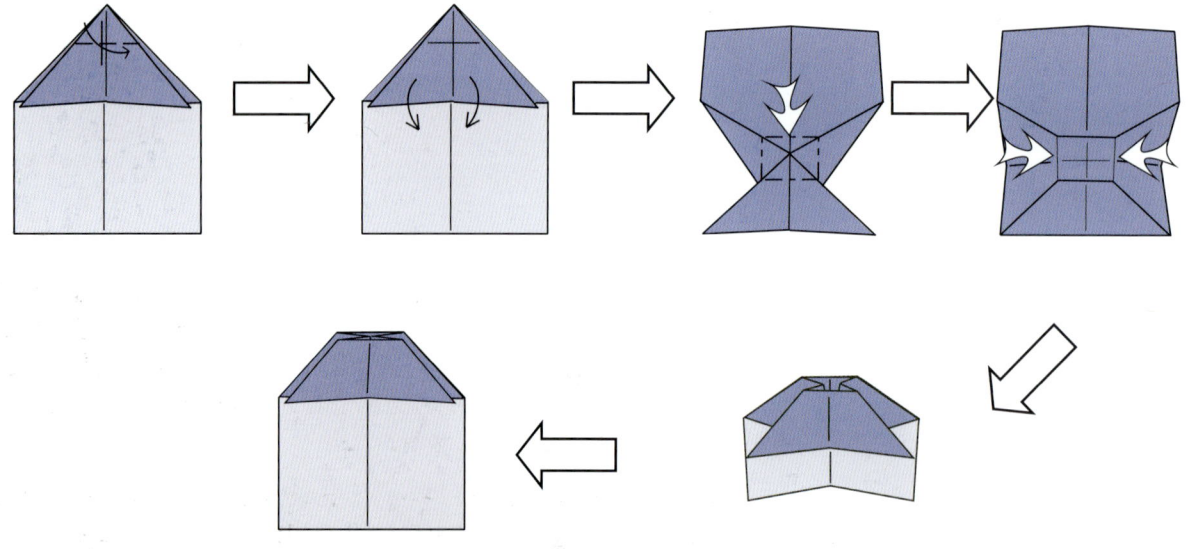

To invent an airplane is nothing. To build one is something. To fly is everything. — Otto Lilienthal

Getting Started

Louis Bleriot

We all know that things were not always as they are now. There once was a time when airplanes did not soar through the sky. And there was once a time when courageous and inventive people first took to the skies in machines they'd created themselves.

One such man was Louis Bleriot, who is remembered for a number of firsts in aviation. He was the first man ever to fly across the English Channel, invented the stick-and-rudder control system, and moved the airplane toward its modern form, with an engine in front and rearward rudder and elevators.

Bleriot could easily have coined the saying *to make a small fortune in aviation, start with a larger fortune*. He had amassed a great deal of money manufacturing acetylene headlights for automobiles in the early days of automotive transport, and he married an heiress, but he expended his capital on experiments to create a powered aircraft.

The Bleriot XI

Unfortunately, his experiments all ended in failure when he crashed his early and very dangerous designs. After seeing the Wright brothers' aircraft at an exhibition in Paris, he realized the importance of roll control, and incorporated that into his eleventh design. The Bleriot XI was successful, and it quickly set records for cross-country flight.

At the time, the *London Daily Mail* was offering a prize of 1,000 pounds for the first person to fly across the English Channel. Although he had competitors, Bleriot won the prize on 25 July 1909, when he piloted his model XI from Calais, crashing onto the shores of Dover. This was a good thing, since Bleriot's finances were at a precarious point.

This accomplishment was as famous and celebrated in its day as Lindbergh's later trip across the Atlantic. For a time the Bleriot XI was the most commonly manufactured airplane in the world, and set the pattern for most aircraft to come. Bleriot stayed active in aviation until his death in 1936.

Getting Started

Smart Dart

A simple airplane that will fly fast or slow, depending on the whim of its creator.

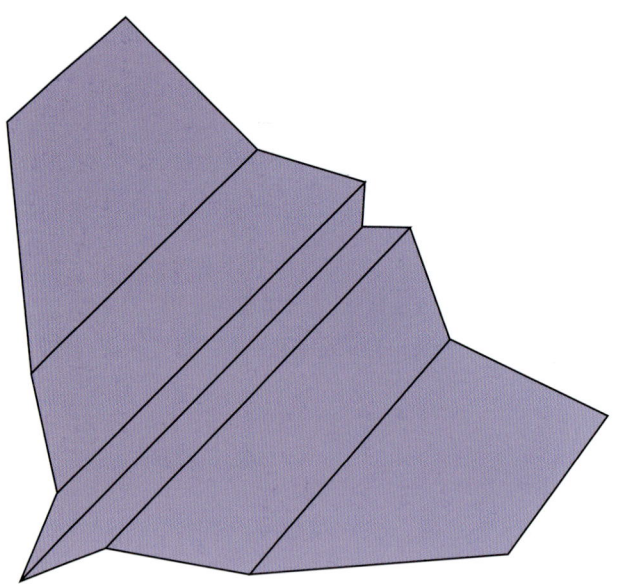

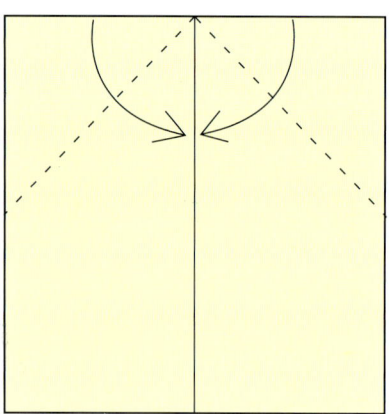

1. Begin with a square creased in half lengthwise. Valley fold the corners so that the top edges lie on the center.

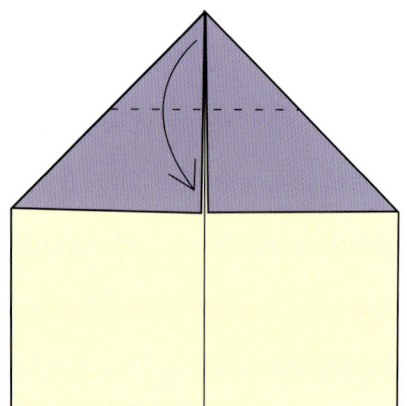

2. Fold the top point down to the bottom of the colored triangle.

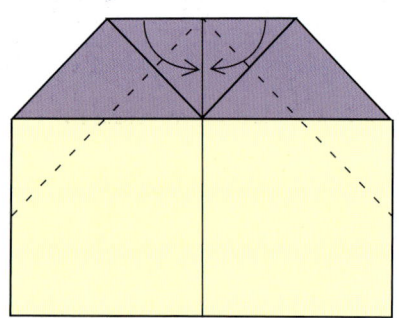

3. Valley fold so that the top edges again lie on the center.

4. Turn over.

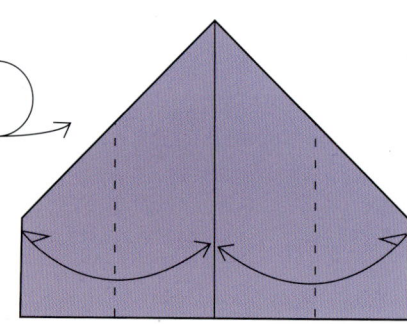

5. Valley fold the sides into the center. Crease well, unfold, and turn back over.

Smart Dart

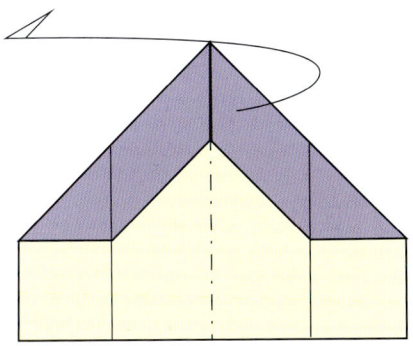

6. Mountain fold the embryonic dart in half.

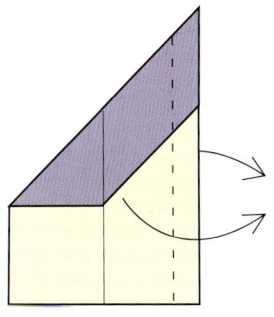

7. Valley fold the wings down.

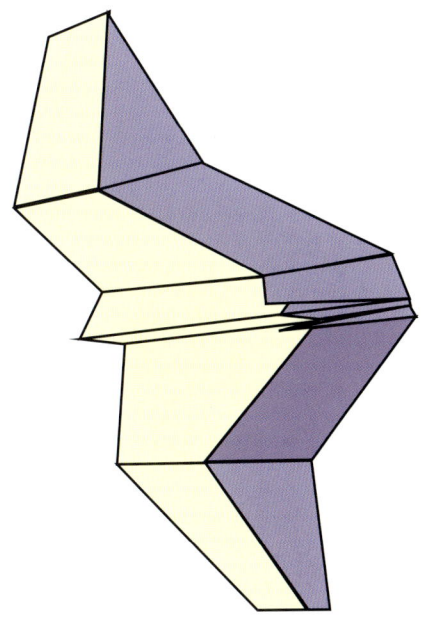

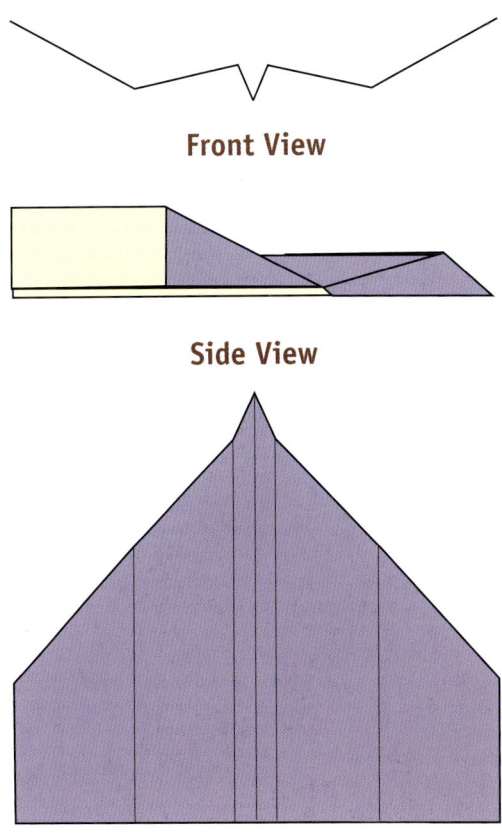

Front View

Side View

Top View

With its forward center of gravity and swept-back wings, Smart Dart can fly very rapidly. Give it a hard throw, and see how far you can get it to go.

Aeronautics was neither an industry nor a science. It was a miracle. — *Igor Sikorsky*

Getting Started

Wild One

A truly crazy stunt plane that's as fun to fly as it is easy to fold. Start with step 3 of the Smart Dart.

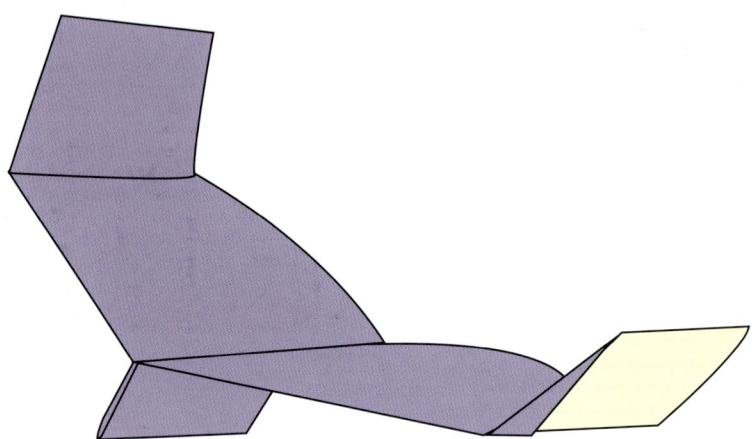

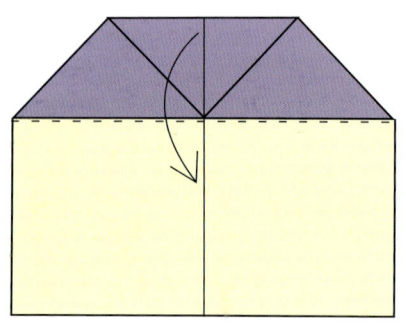

1. Valley fold straight across the front.

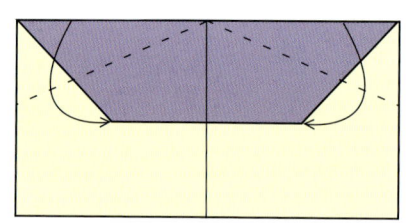

2. Valley fold so that the folded edge on top touches the point shown.

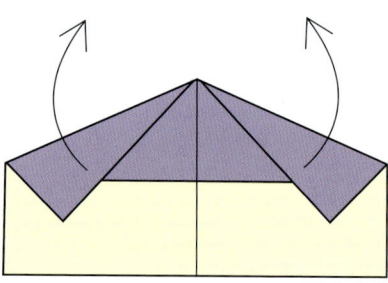

3. Unfold.

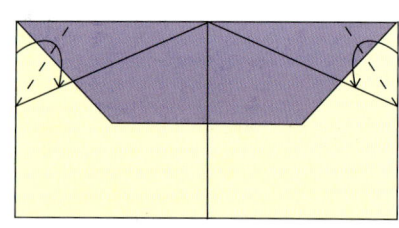

4. Valley fold so that the raw outside edge lies on the crease made in step 2.

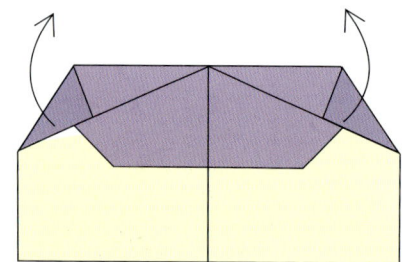

5. Unfold.

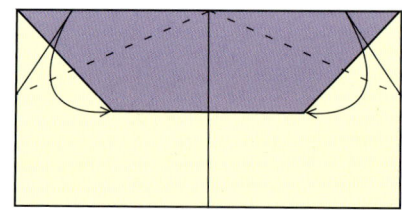

6. Valley fold on the crease made in step 2.

The engine is the heart of an aeroplane, but the pilot is its soul. — Sir Walter Alexander Raleigh

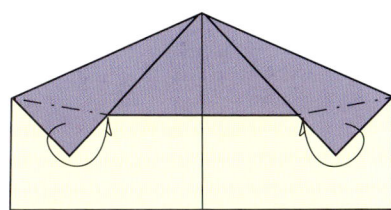

7. Mountain fold along the creases made in step 4. Tuck the flaps behind all the other layers, to lock up the front.

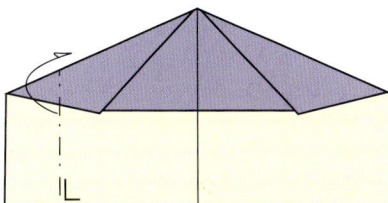

8. Mountain fold parallel to the center and perpendicular to the back. You can feel where the fold should be, because it's where the paper gets thicker.

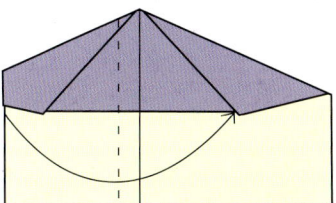

9. Valley fold the wing by bringing the folded edge over to the point shown.

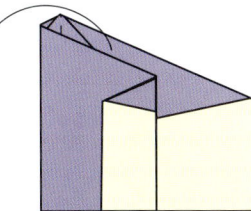

10. Mountain fold along the centerline.

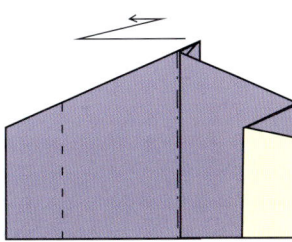

11. Fold the other wing so that it's the same as the first.

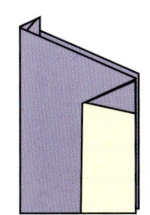

12. Wild one!

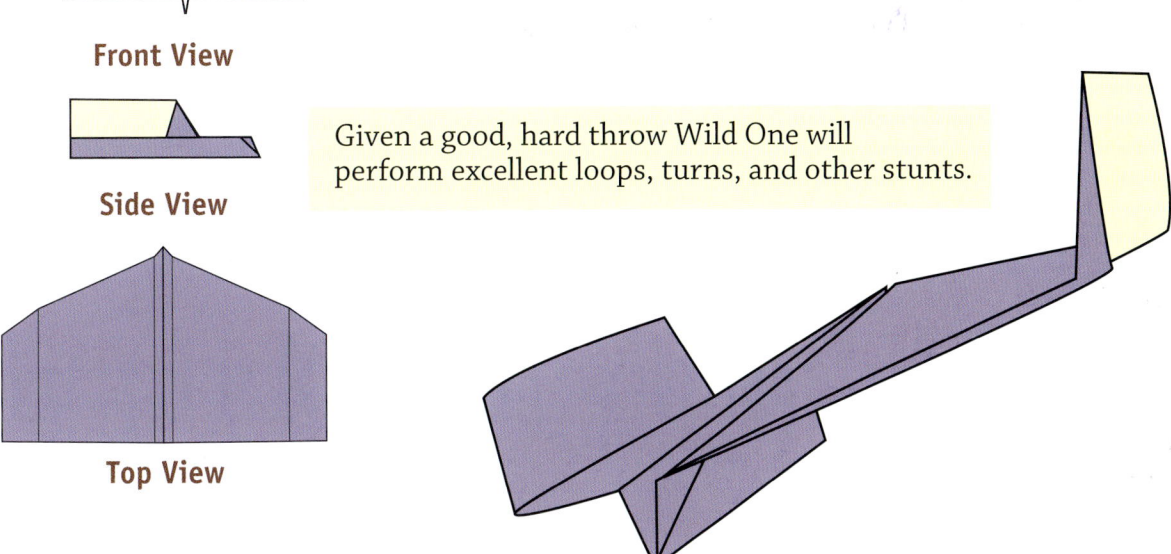

Front View

Side View

Top View

Given a good, hard throw Wild One will perform excellent loops, turns, and other stunts.

Both optimists and pessimists contribute to society. The optimist invents the aeroplane, the pessimist the parachute. — *George Bernard Shaw*

Getting Started

Protoharrier

I once created an airplane that I called Harrier because of the use of rabbit ears in its folding process. It was not named for the British VTOL (vertical takeoff and landing) aircraft, despite the fact that it went to England. This is a somewhat simpler version. Begin with a square of kami creased down the middle, white-side up.

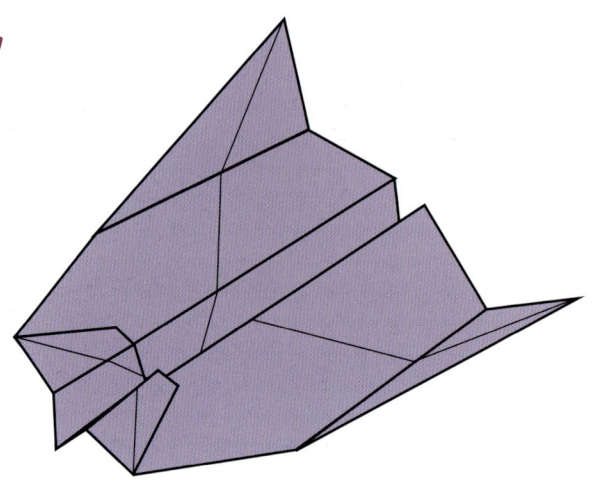

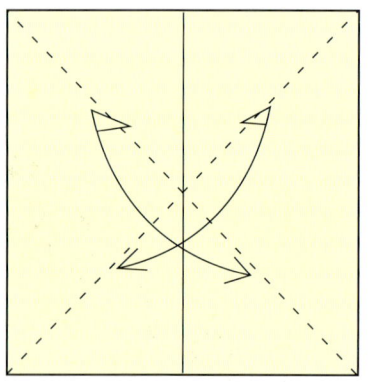

1. Crease both diagonals.

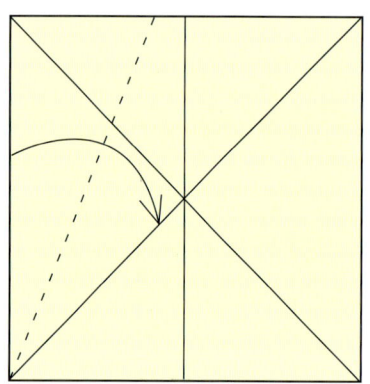

2. Valley fold the raw edge to the crease you just made.

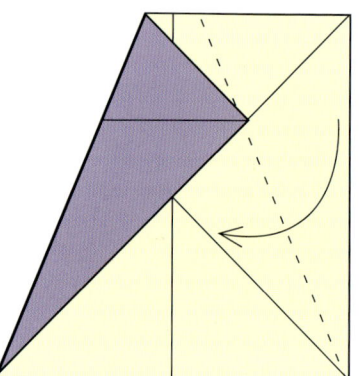

3. Like so. Repeat on the other side.

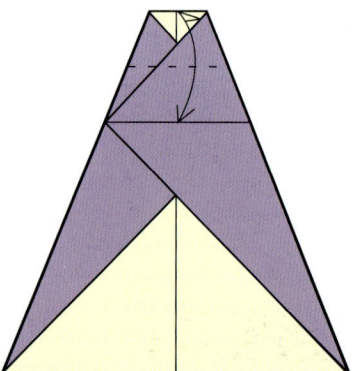

4. Valley fold so that the top edge lies on the crease shown. Crease well and unfold.

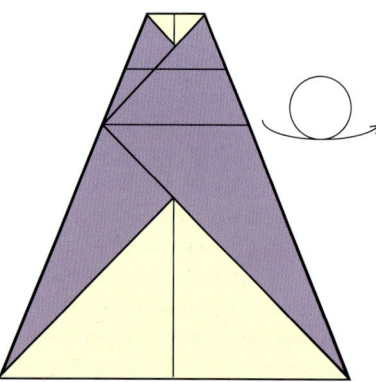

5. Turn over.

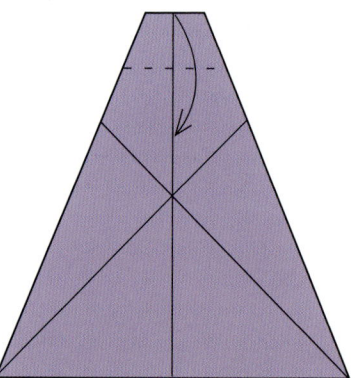

6. Valley fold along the crease you made in step 4.

26

Protoharrier

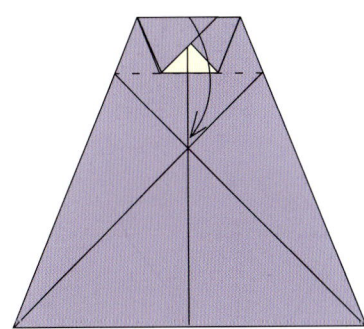

7. Valley fold the top section down. There will be a crease on the back that you can use as a guide.

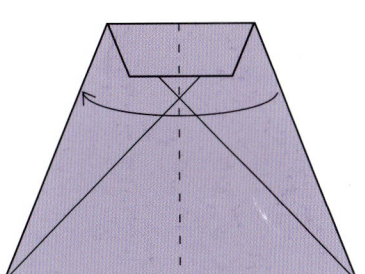

8. Valley fold the Protoharrier in half.

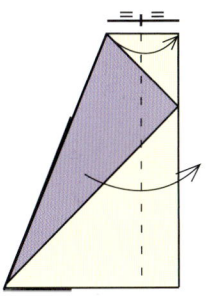

9. Valley fold the wings down. The crease should run straight up and down, and cut the front edge in half.

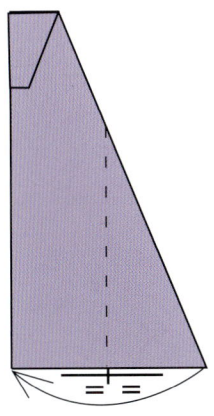

10. Valley fold the stabilizers. Again, the crease runs straight up and down, bisecting the rear edge.

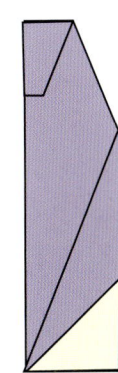

Protoharrier complete

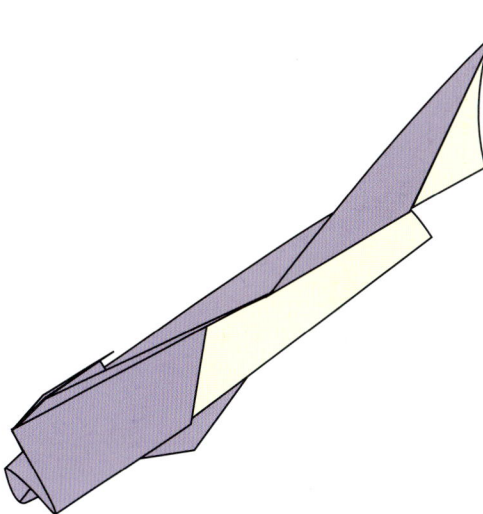

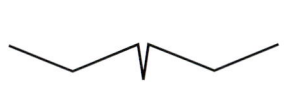

Front View

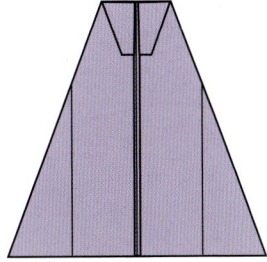

Top View

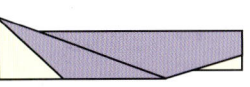

Side View

27

Getting Started

Nakamura Canard

A book that had a great influence on me was *Flying Origami* by Eiji Nakamura. It had canards and futuristic airplanes galore, and it was very, very cool. I've reconfigured the canard from this collection to be made from a square.

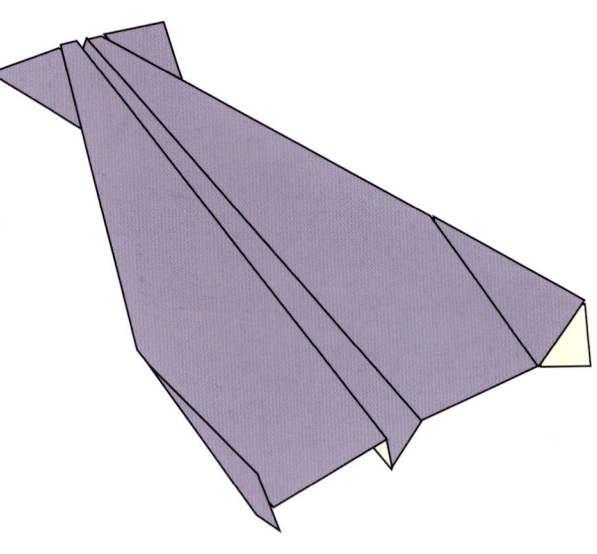

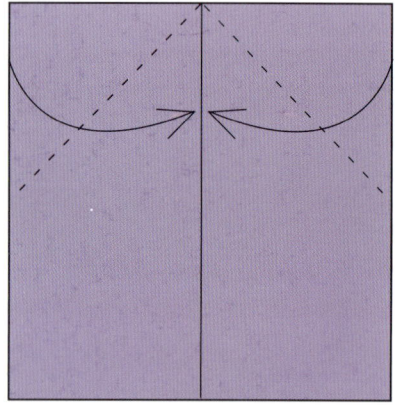

1. Valley fold from the top corners to the center.

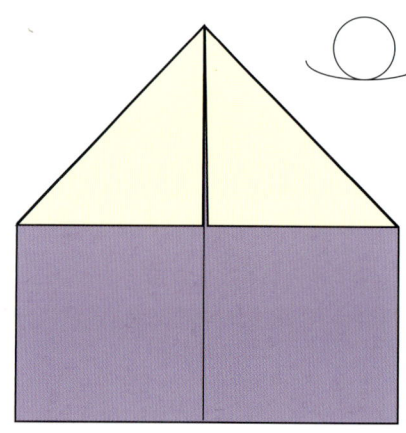

2. Turn over.

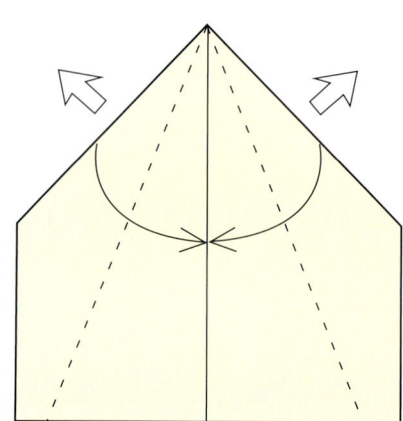

3. Valley fold so that the folded edges at the top lie on the centerline. The flaps in the back should swing out.

To most people, the sky is the limit. To those who love aviation, the sky is home. — Anonymous

28

Nakamura Canard

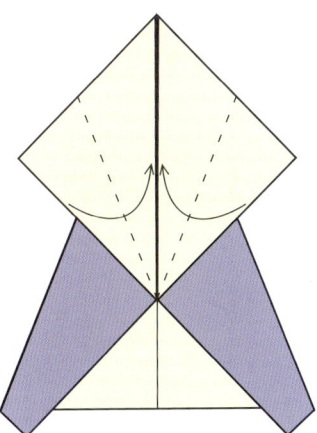

4. Valley fold so that the raw edges on the bottom of the square lie on the center.

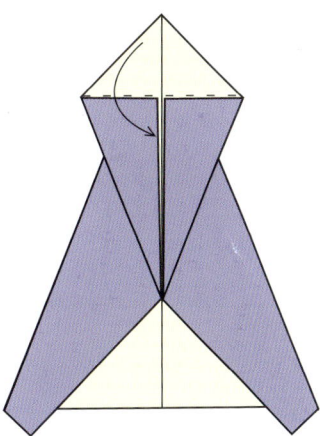

5. Valley fold the triangle at the top down.

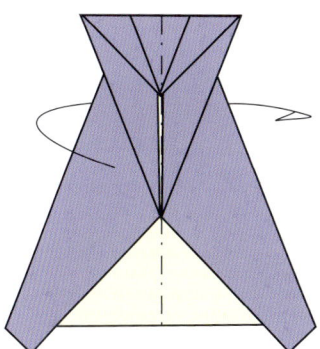

6. Mountain fold the airplane in half.

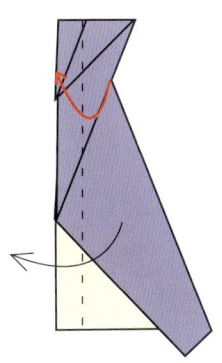

7. Valley fold the wings so that the point shown in red touches the bottom.

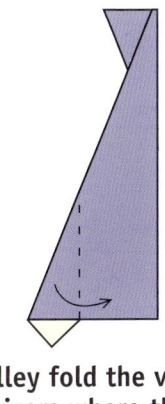

8. Valley fold the vertical stabilizers where the layers meet.

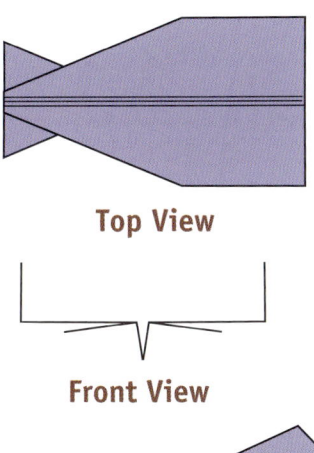

Top View

Front View

Side View

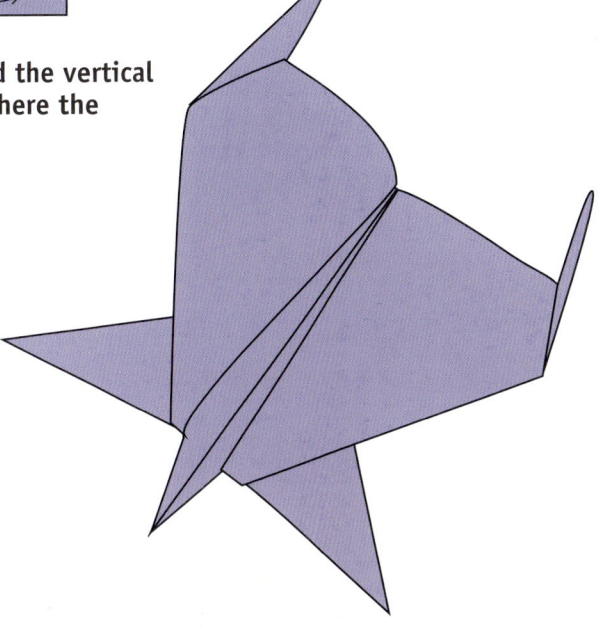

29

Getting Started

Sharkie

Our first airplane made on the diagonal. The front looks like a shark, though it performs more like a flying fish. This felt more like it was discovered than invented. Start with a square creased down the diagonal.

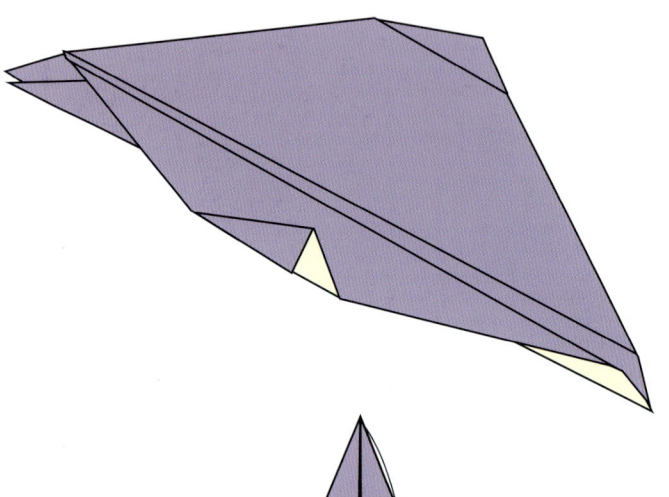

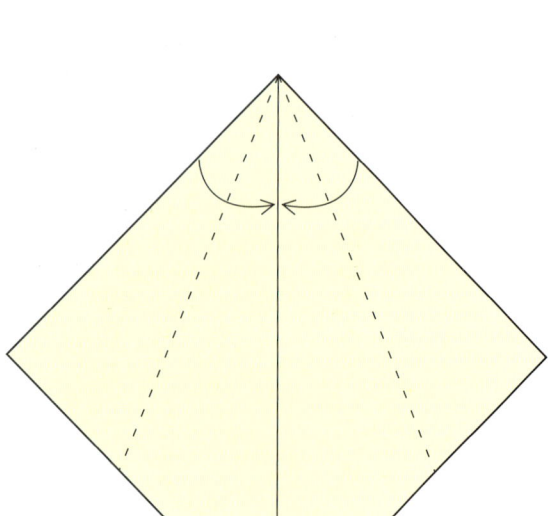

1. Valley fold so that the two raw edges on the top meet the centerline. In origami parlance this is called a Kite Fold.

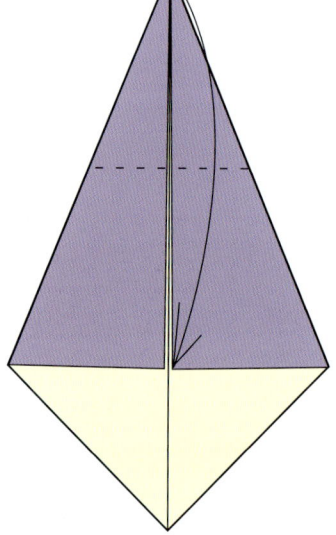

2. Valley fold the top point to the bottom of the colored flaps.

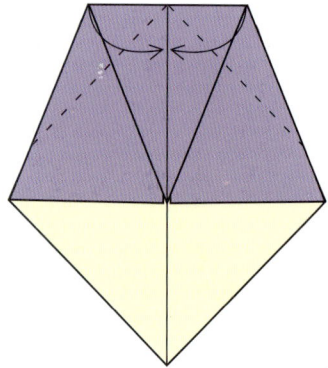

3. Valley fold the top folded edges into the center.

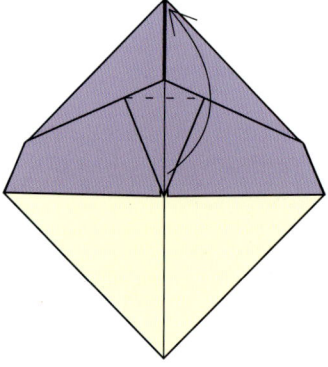

4. Valley fold the middle flap up to cover the ones above it.

Sharkie

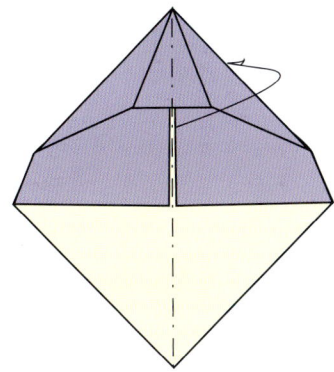

5. Mountain fold in half.

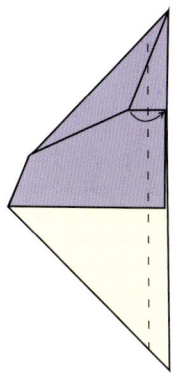

6. Valley fold the wings, splitting Sharkie's mouth in half.

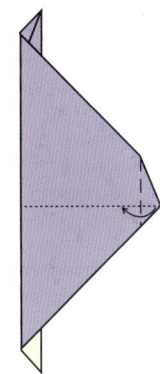

7. Valley fold the ends of the wings to make vertical stabilizers. You may want to hold Sharkie up to a light so that you can see the layer underneath, which will be your guide in this fold (the point at the end will line up with the edge of the darker layer).

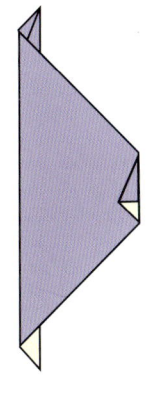

8. You've completed the Sharkie! Open up the wings perpendicular to the fuselage, and the stabilizers perpendicular to the wings.

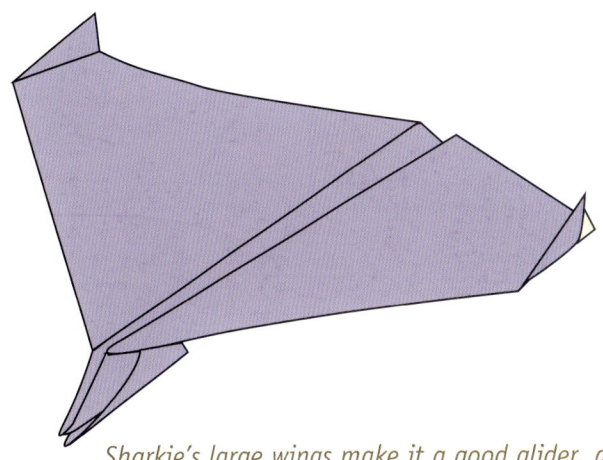

Sharkie's large wings make it a good glider, as its weight is spread out a great deal. Its wings are swept back, however, and it will give an interesting flight if you throw it fast.

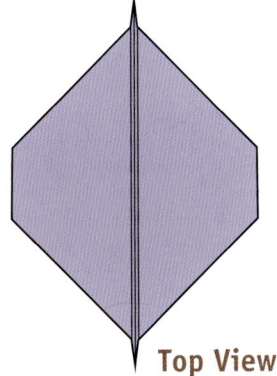

Top View

Front View

Side View

31

Getting Started

Sciaetta

This was the first of the kami airplanes, and remains one of the best. Begin with a square, white-side up and creased down the diagonal.

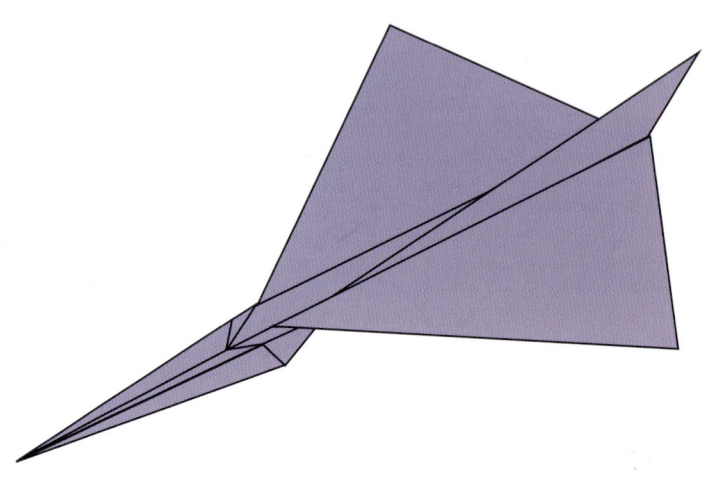

1. Valley fold so that the two raw edges on the top meet the centerline.

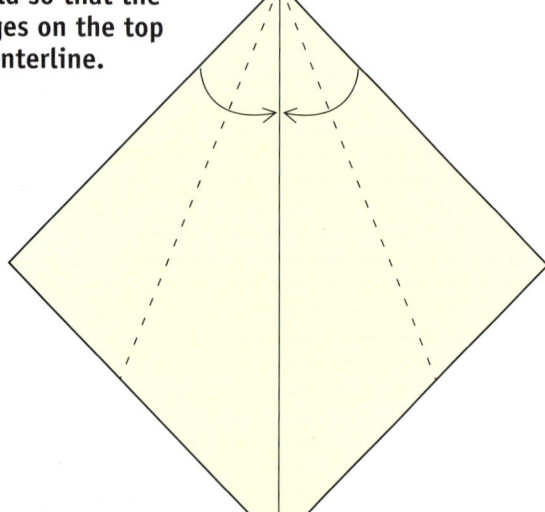

2. Unfold.

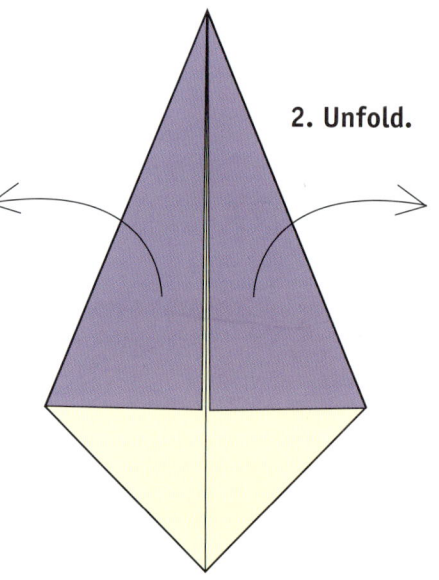

3. Valley fold so that the lower part of the raw edge meets the crease you just made.

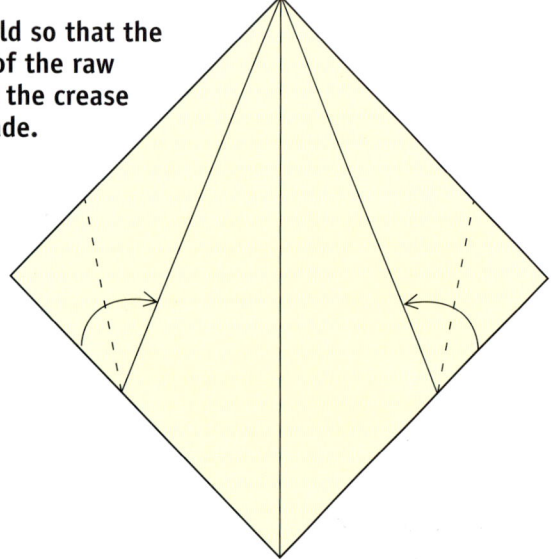

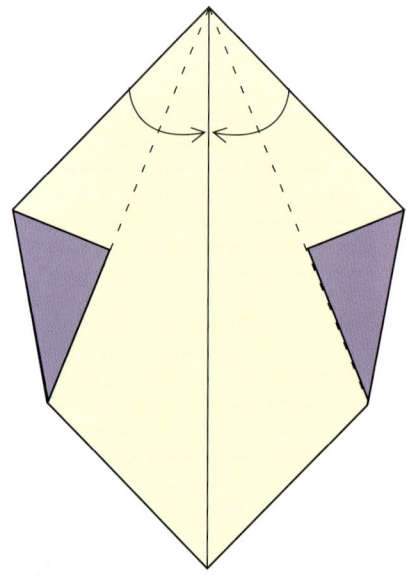

4. Valley fold the sides back in.

32

Sciaetta

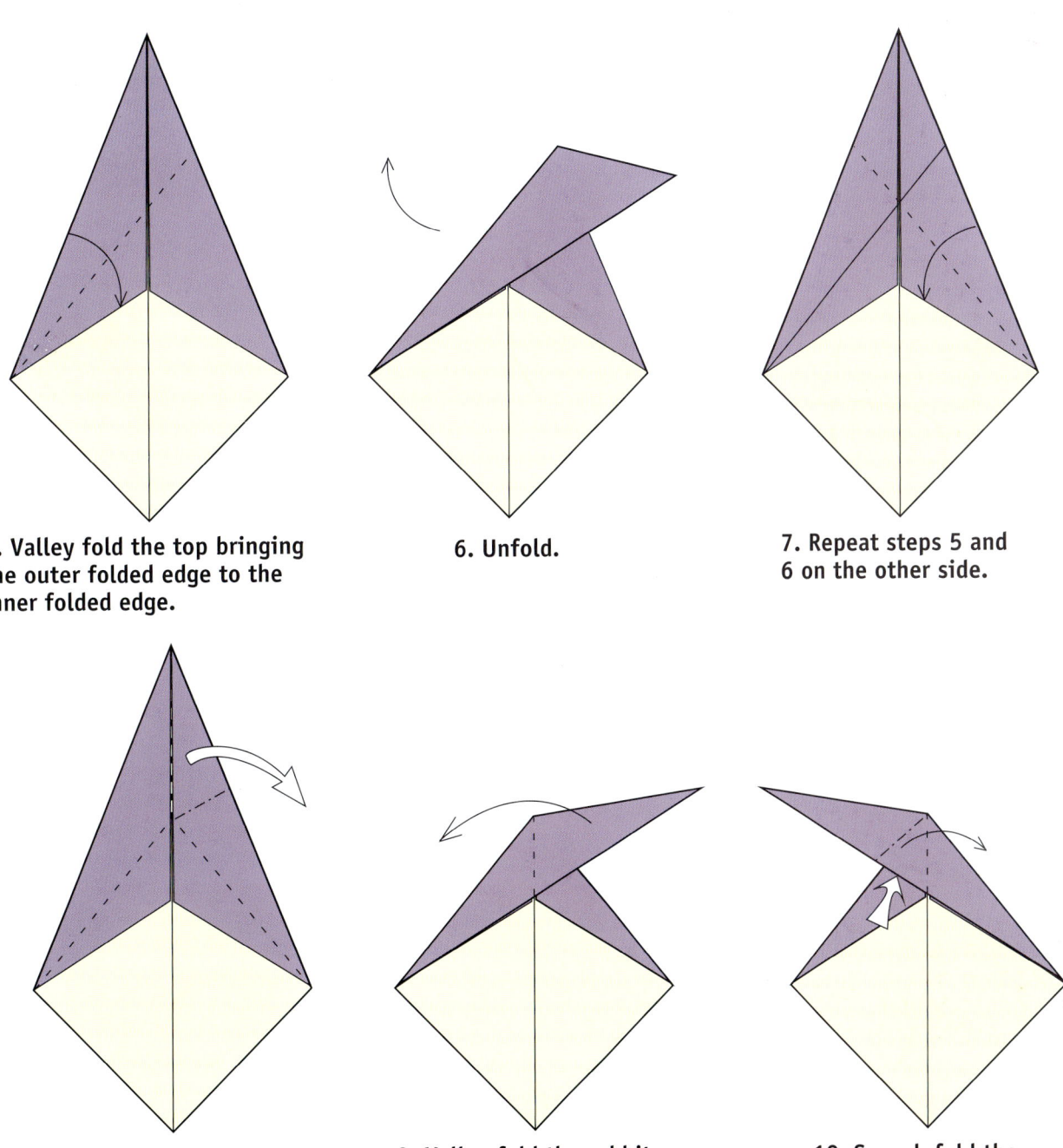

5. Valley fold the top bringing the outer folded edge to the inner folded edge.

6. Unfold.

7. Repeat steps 5 and 6 on the other side.

8. Rabbit ear the top to the right, and crease well.

9. Valley fold the rabbit ear to the other side and crease it well.

10. Squash fold the upper flap.

The Wright brothers flew through the smoke screen of impossibility. — Dorothea Brande

Getting Started

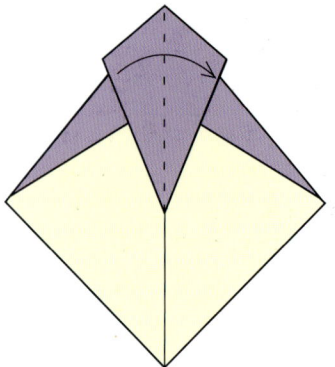

11. Valley fold the upper diamond along its midline.

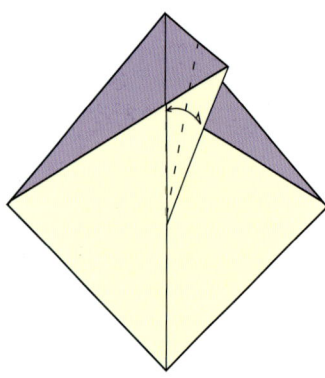

12. Valley fold the upper diamond bisecting the lower angle.

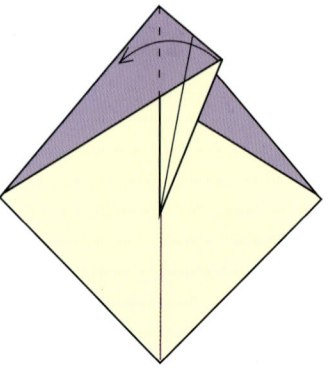

13. Valley fold the upper diamond back to the left.

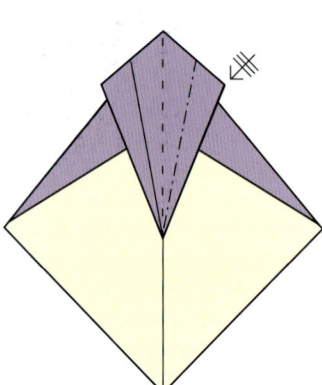

14. Repeat steps 11–13 on the other side.

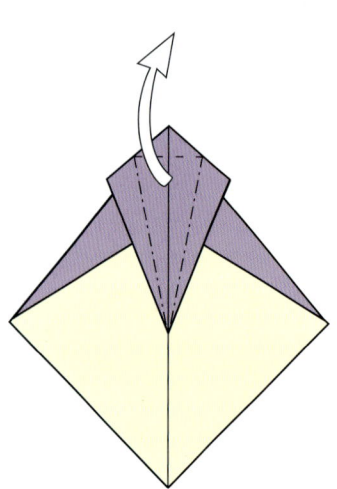

15. Petal fold upwards.

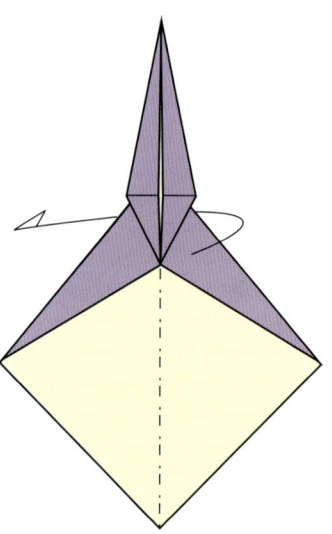

16. Mountain fold the airplane in half and rotate 90°.

The bluebird carries the sky on his back. — Henry David Thoreau

Sciaetta

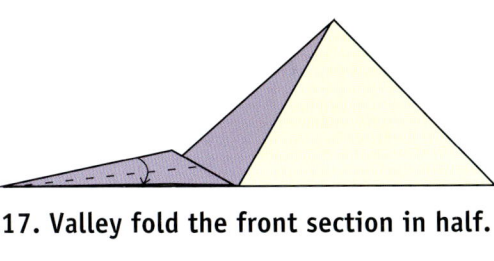

17. Valley fold the front section in half.

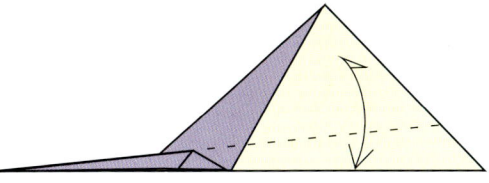

18. Valley fold the wings along the same angle as the front. Crease and unfold.

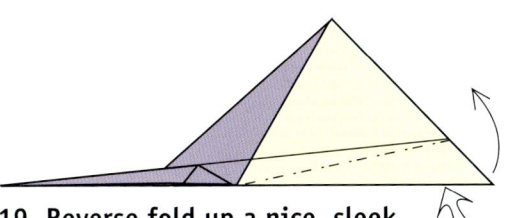

19. Reverse fold up a nice, sleek tail as far as you can.

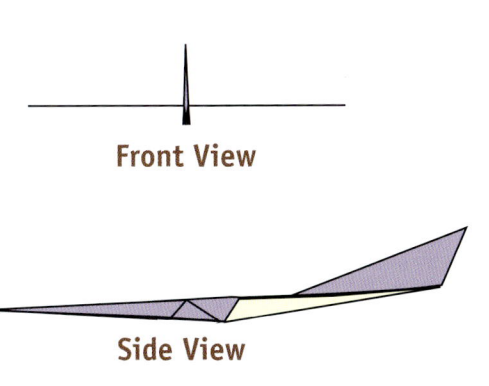

Front View

Side View

Top View

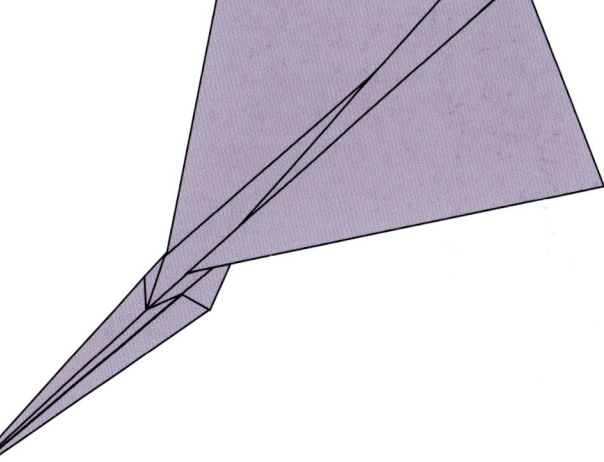

Given a hard throw, Sciaetta gives awesome, fast flights. Its sleek design cuts through the air, and boy does it look good!

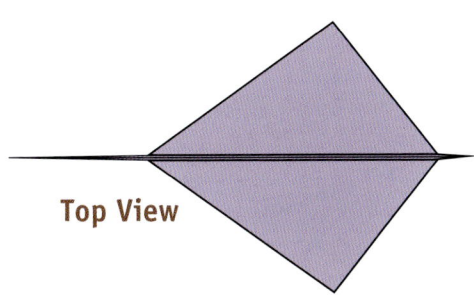

35

A Few Good Darts

Clarence Leonard "Kelly" Johnson

Many of our favorite modern aircraft are streamlined and fast. Various current fighter jets bring this to mind. You might be surprised, however, to discover that the fastest jets ever were constructed over four decades ago. No one would be surprised to discover that they were designed by the great Kelly Johnson.

Clarence Leonard "Kelly" Johnson was born in Michigan in 1910 to Swedish immigrant parents. He worked his way through college, applying to Lockheed for a job in 1932. Initially turned down, he returned a year later with a master's degree and was hired. He first came to the attention of Lockheed management when he detected and repaired directional stability problems in its sleek twin-engine Electra airliner, the aircraft Amelia Earhart flew on her ill-fated round-the-world flight.* He quickly advanced at Lockheed, becoming chief research engineer in 1938. In 1939 he single-handedly designed the Lockheed Hudson Bomber in a hotel room in Great Britain. In 1952 he became chief engineer at Lockheed's Burbank California facility, and vice president for research and development in 1956.

Johnson was an incredible designer, creating such advanced aircraft as the P38 Lightning, the F104 Starfighter, the high-altitude U2 (in which Francis Gary Powers was famously shot down in 1956), and the spectacular SR71 Blackbird, which holds the records for both altitude and speed for a jet aircraft. What makes this feat even more amazing is that it was done in the absence of computers or advanced composite technologies. Johnson is credited with the design of over forty aircraft, but his organizational genius is even more legendary. He was able to direct the design and construction of advanced prototypes with astounding speed and efficiency, doing in months what now takes years.

He set up the Lockheed facility at Groom Lake, commonly referred to as "Area 51," and served on the Lockheed board of directors from 1964 until his retirement in 1980. The list of awards he was honored with was quite long. He died of a lengthy illness in 1990.

*Amelia Earhart did not make it around the world, disappearing off Howland Island in 1937. The first woman to fly solo around the world was Geraldine "Gerry" Mock, a suburban housewife and mother of three from my hometown of Columbus, Ohio. In 1964 she flew around the world in her family's Cessna 180, named *The Spirit of Columbus*.

The SR71 Blackbird, the fastest jet in history

Needler

Chaos theory is an area of mathematics with profound implications. Its application suggests that folding and launching this airplane could change the weather in Greenland. You should therefore fold it, since you're not going to hurt the weather in Greenland any.

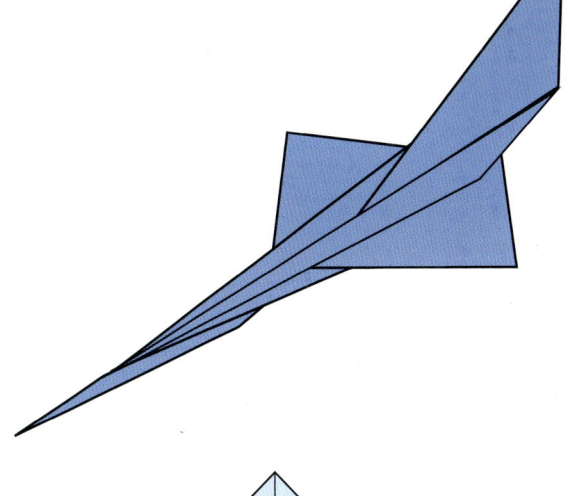

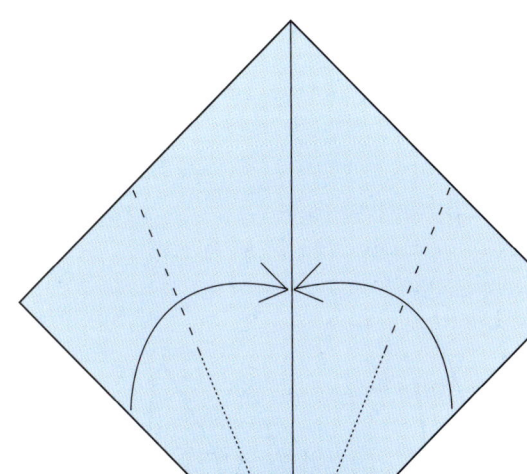

1. Kite fold on both sides, but crease only the upper half of the fold.

2. Valley fold the top down where the creases made in step 1 meet the outside edges.

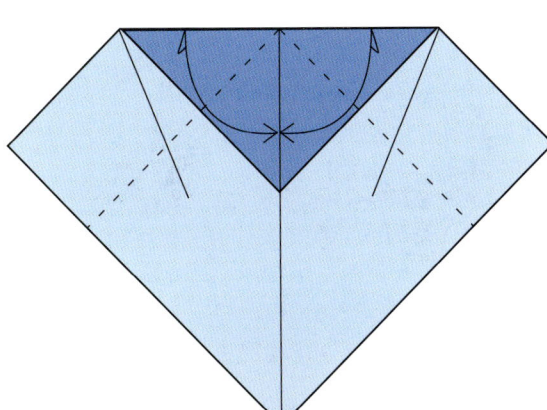

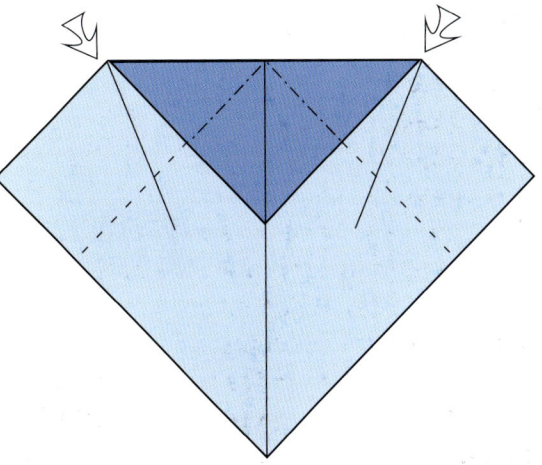

3. Valley fold the sides down at a 45° angle. Do this by lining up the folded edges at the top and the centerline.

4. Reverse fold along the creases made in step 3.

37

A Few Good Darts

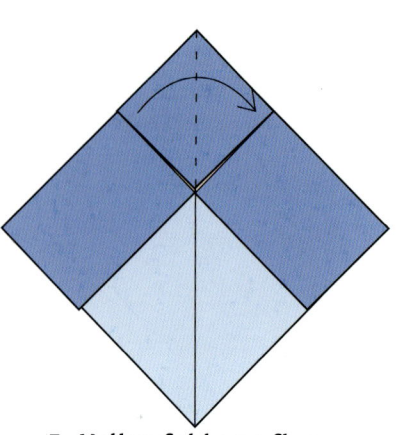

5. Valley fold one flap to the other side.

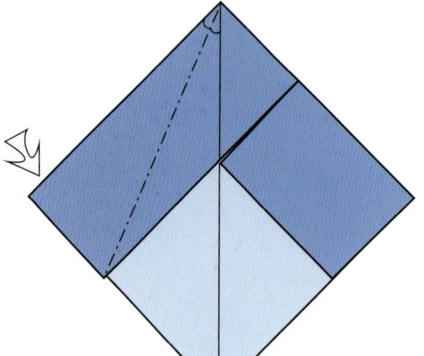

6. Reverse fold the flap in half.

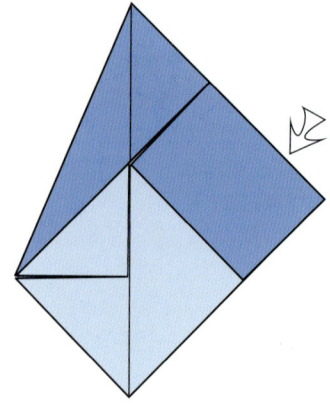

7. Repeat steps 5 and 6 on the other side.

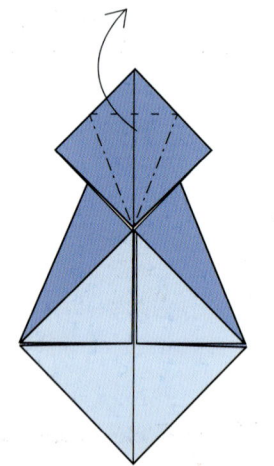

8. Petal fold the square flap upward, then turn over.

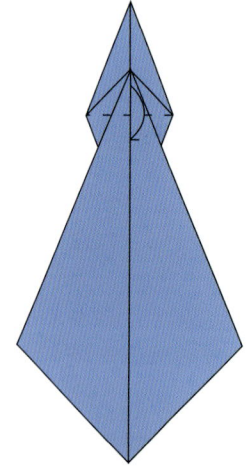

9. Valley fold the flap down, and turn back over.

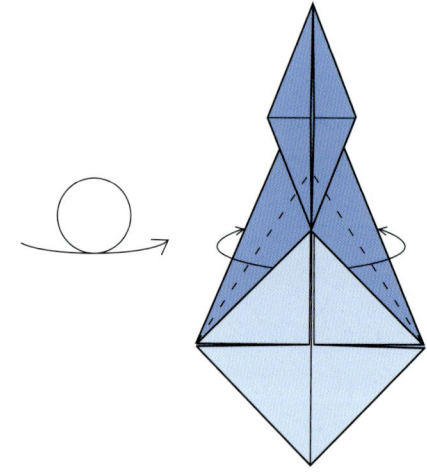

10. Valley fold the upper layer of the long flaps in half.

11. Reverse fold the wings outward. The fold will run along the light colored flaps, and the crease at the midline will line up with the outside edge of the fuselage.

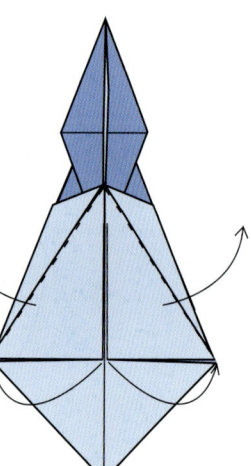

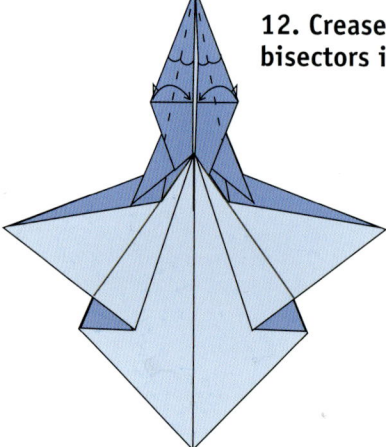

12. Crease the angle bisectors in the front.

38

Needler

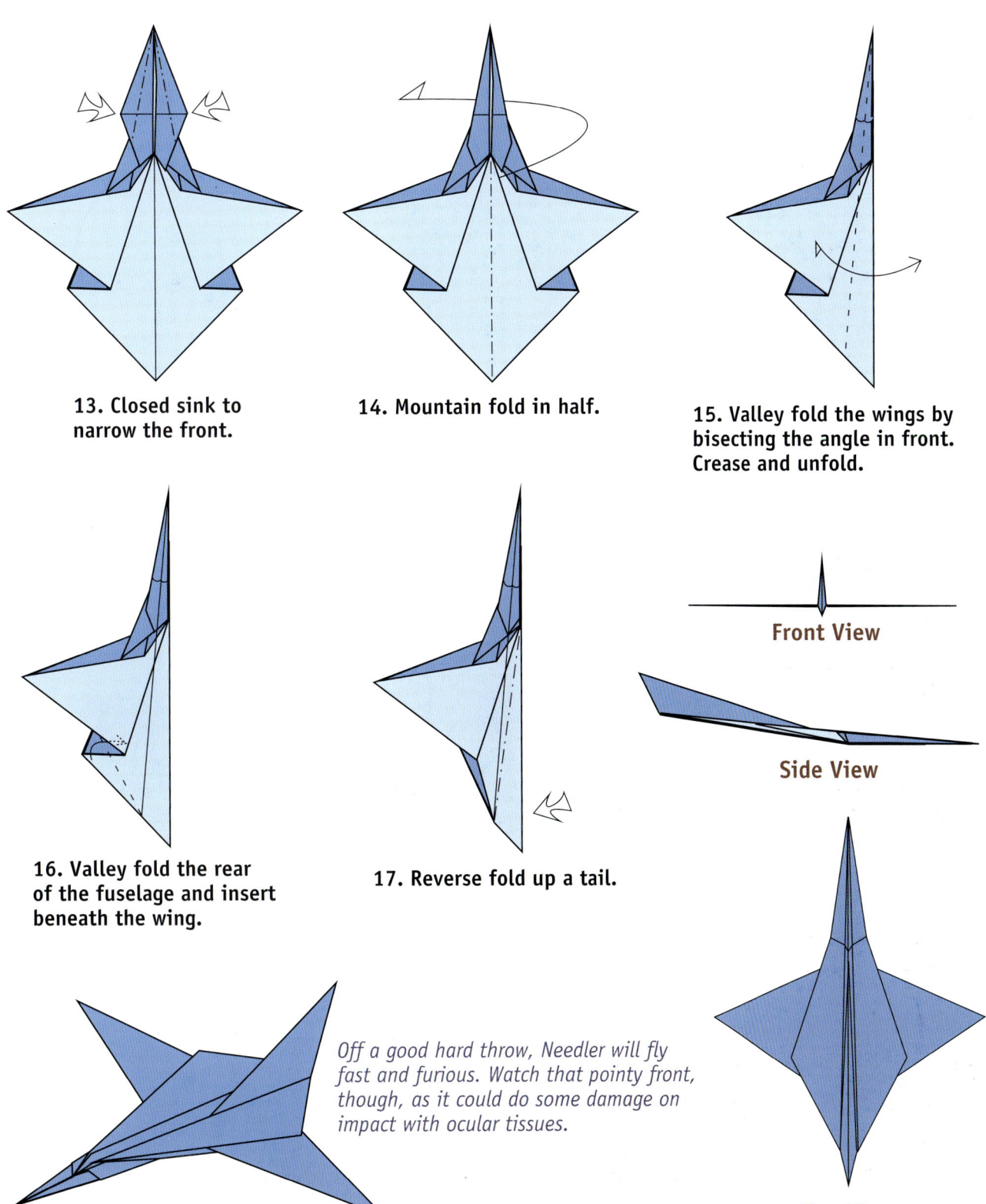

13. Closed sink to narrow the front.

14. Mountain fold in half.

15. Valley fold the wings by bisecting the angle in front. Crease and unfold.

16. Valley fold the rear of the fuselage and insert beneath the wing.

17. Reverse fold up a tail.

Front View

Side View

Top View

Off a good hard throw, Needler will fly fast and furious. Watch that pointy front, though, as it could do some damage on impact with ocular tissues.

A Few Good Darts

The Un-unfoldable Airplane

Once upon a time the origami genius John Cunliffe came up with a box that was easier to fold than unfold. Naturally, one needs a paper airplane with such properties.

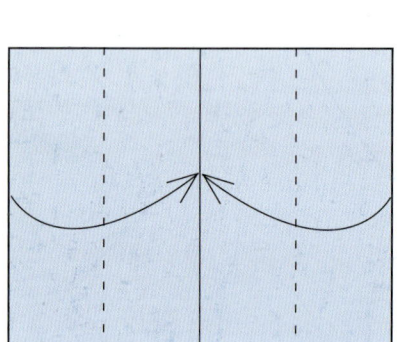

1. Cupboard fold on both sides.

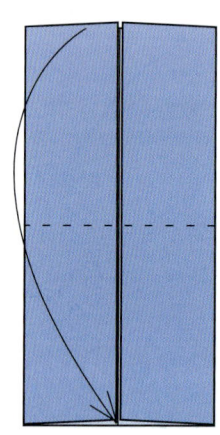

2. Valley fold in half lengthwise.

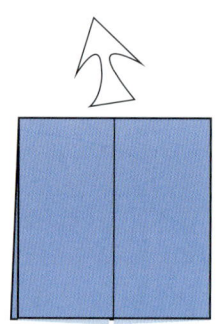

3. Unfold to step 1.

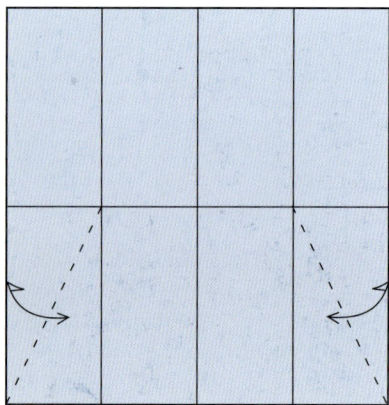

4. Valley fold and unfold from the corner to the intersection shown.

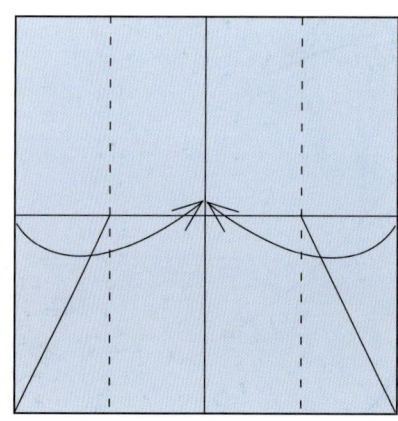

5. Reform the cupboard fold.

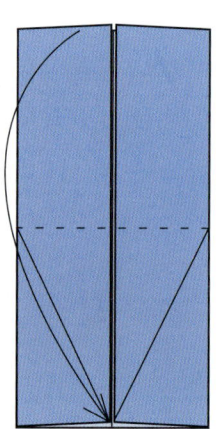

6. Valley fold in half lengthwise.

The Un-unfoldable Airplane

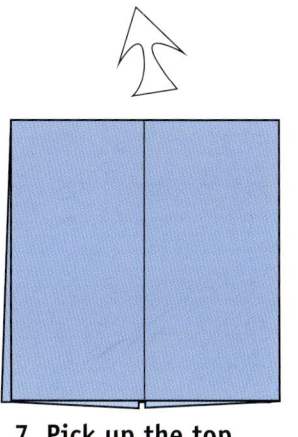

7. Pick up the top layer partway.

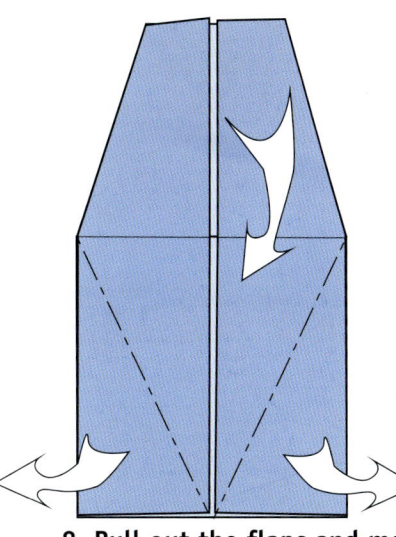

8. Pull out the flaps and mountain fold on the creases made in step 4, while putting top back down. Flatten and crease well.

9. Unfold to step 2.

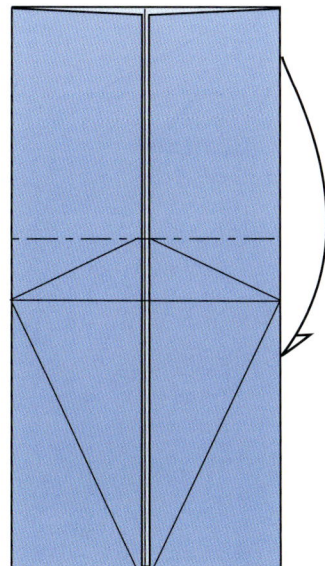

10. Mountain fold behind the intersection of the indicated creases.

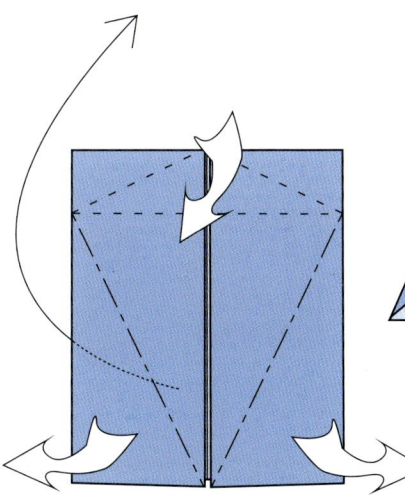

11. Refold along the creases made in step 8, bringing the underlying back up again.

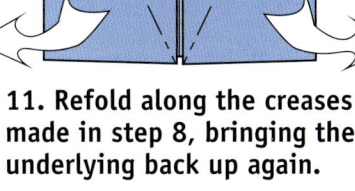

12. Cupboard fold the top layer only. The aircraft will not lie flat.

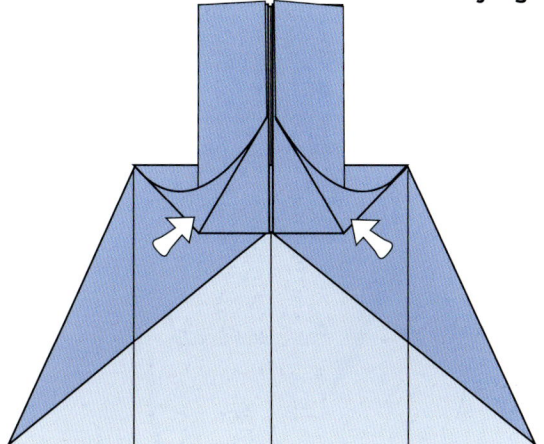

13. Flatten the resulting pockets.

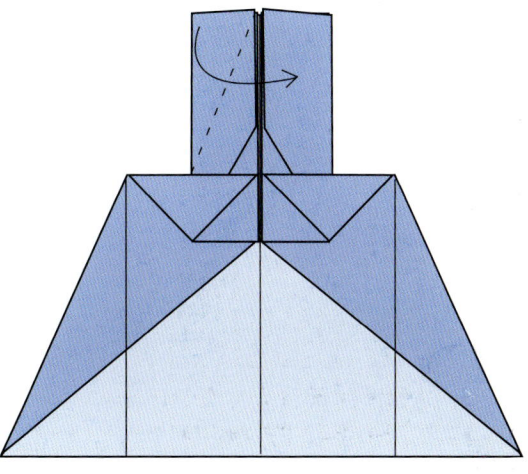

14. Valley fold the top.

A Few Good Darts

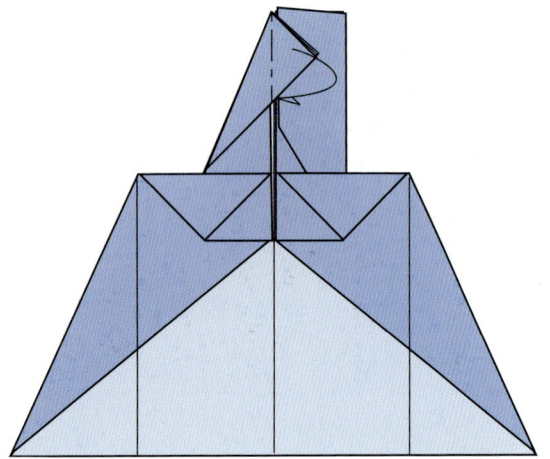

15. Mountain fold the upper flap at the midline, and tuck it under all the layers beneath it. This should lock the front in place.

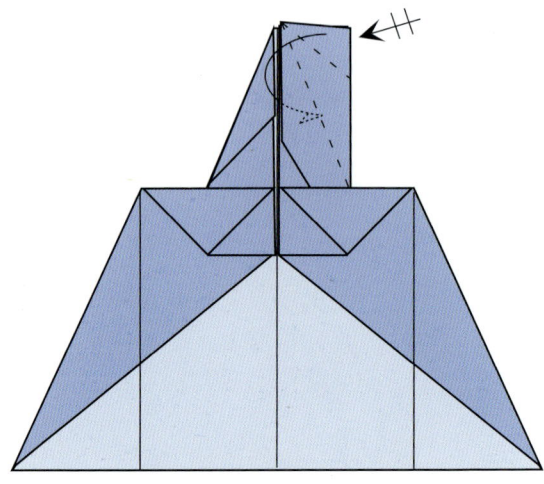

16. Repeat steps 14 and 15 on the other side.

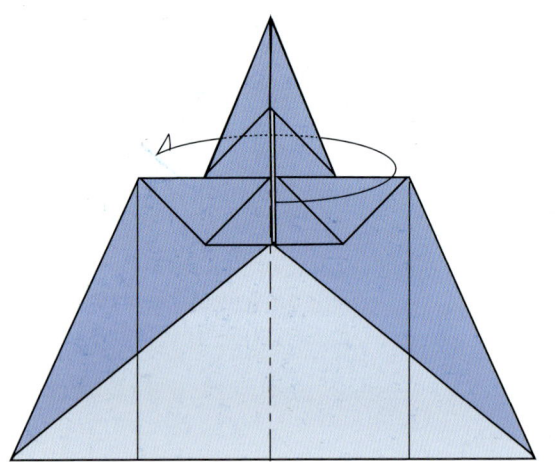

17. Mountain fold the whole thing in half.

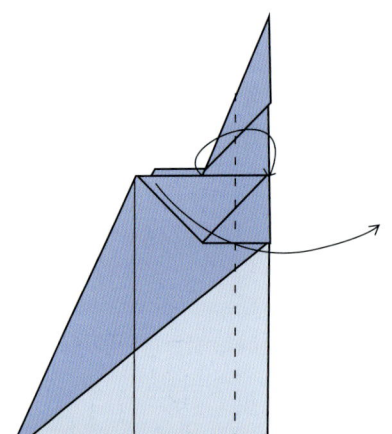

18. Valley fold the wing down by halving the area in front. Repeat behind.

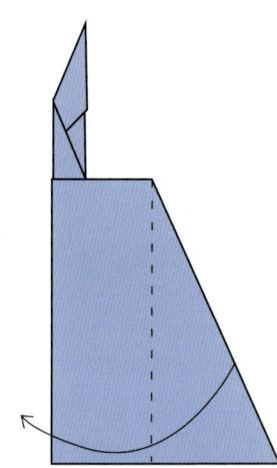

19. Valley fold the stabilizers along a preexisting crease.

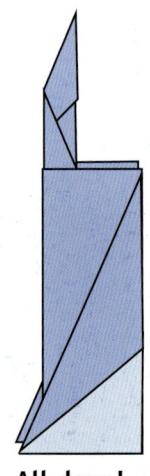

All done!

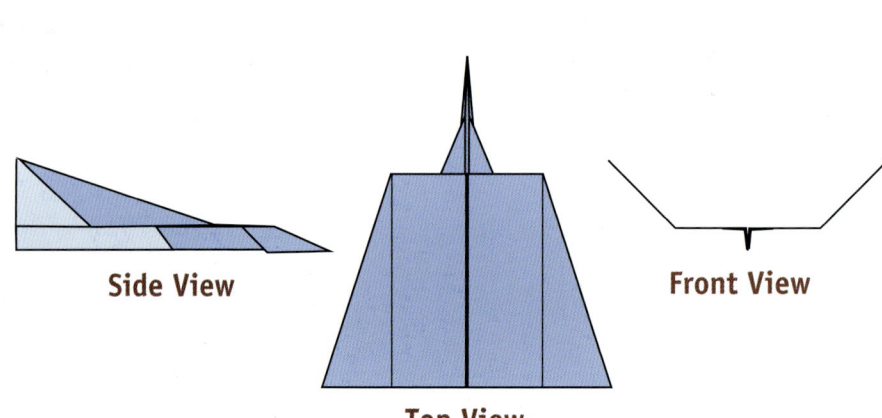

Side View

Top View

Front View

42

Bird-base Fighter

The Bird-base produces four flaps that can be amended into two wings, a head, and a tail. It makes good airplanes, too. Use duo paper if you have it. A light-colored bottom against a dark top makes BBF look like it's got jet engine exhaust coming out above the wing, just like a stealth aircraft. Real stealth jets use such a configuration to keep their exhaust from being seen from the ground by heat-seeking missiles or unfriendly eyes.

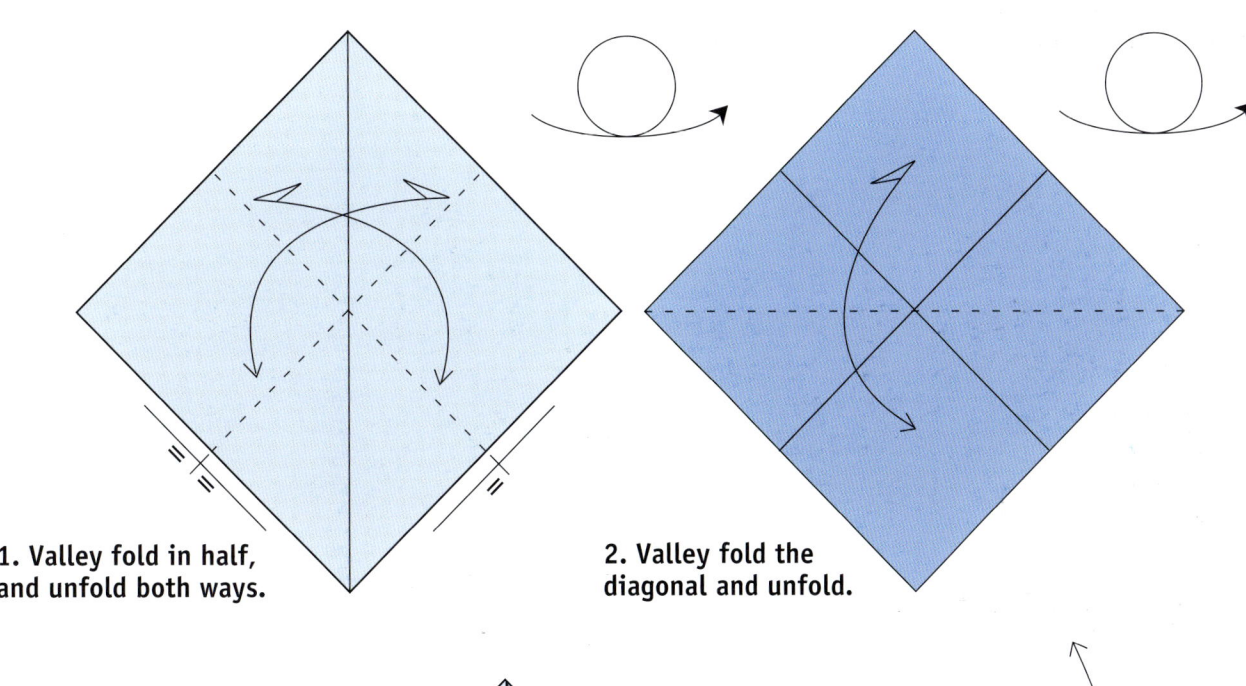

1. Valley fold in half, and unfold both ways.

2. Valley fold the diagonal and unfold.

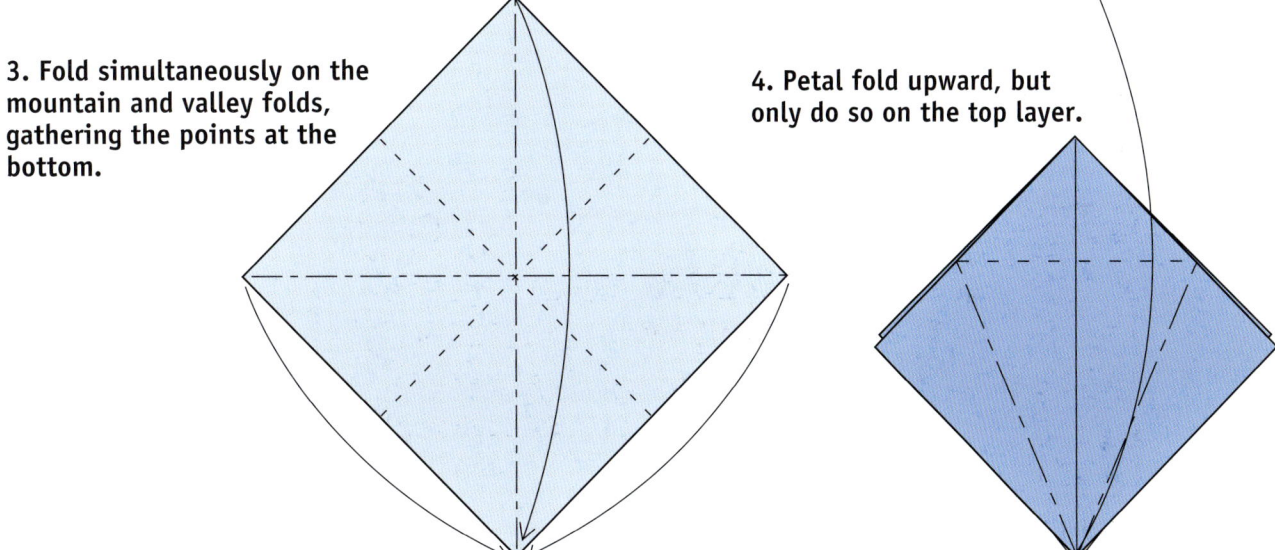

3. Fold simultaneously on the mountain and valley folds, gathering the points at the bottom.

4. Petal fold upward, but only do so on the top layer.

A Few Good Darts

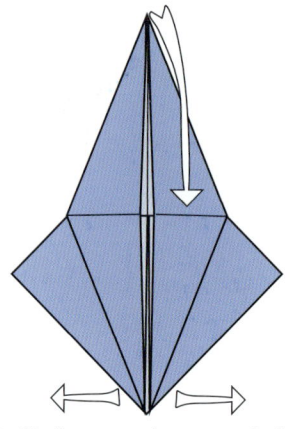

5. Pull the top down and the bottom flaps out to the side.

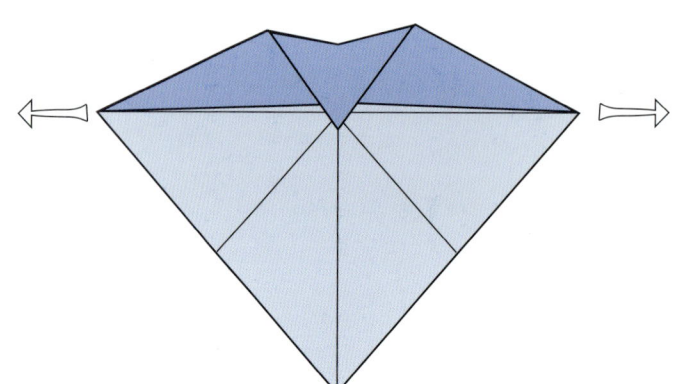

6. (Step 5 in progress.) Keep pulling until the top is straight across, and not at an angle. Then flatten.

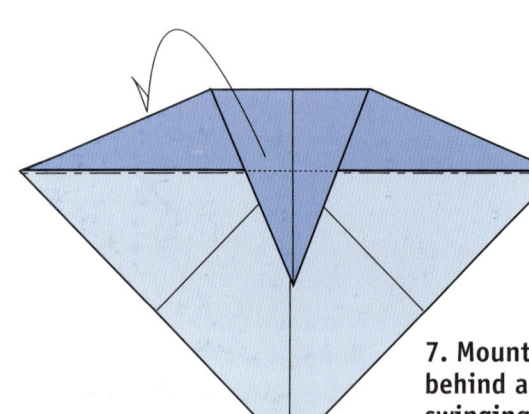

7. Mountain fold the top behind along the diagonal, swinging the point upward.

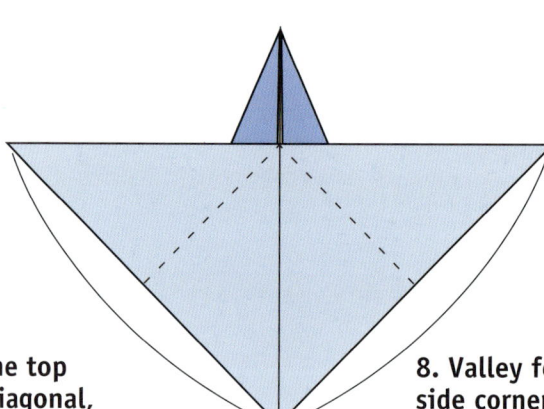

8. Valley fold the side corners down to the bottom point.

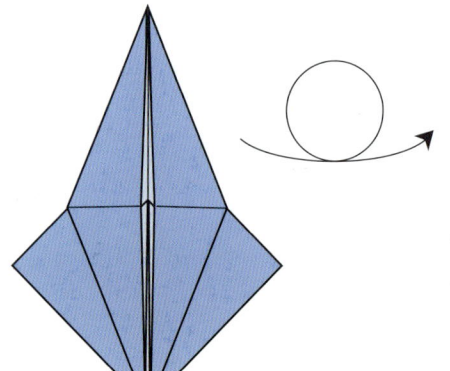

9. Turn over.

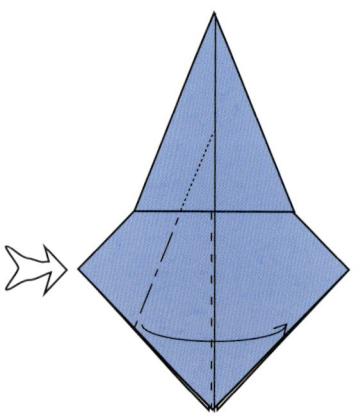

10. Squash fold to the right.

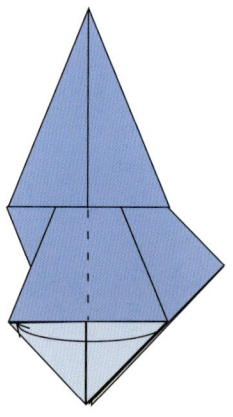

11. Valley fold back to the left.

44

Bird-base Fighter

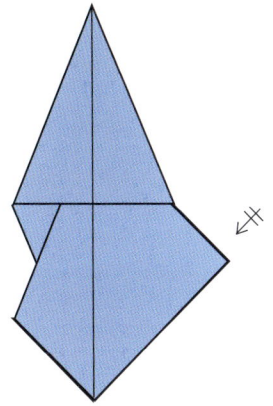

12. Repeat steps 10 and 11 on the other side.

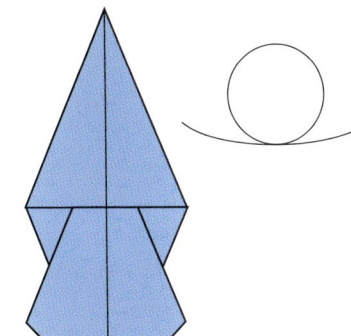

13. Turn over.

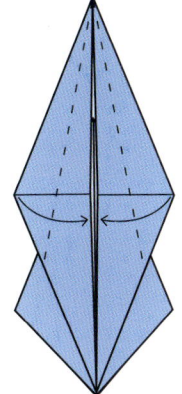

14. Valley fold each side of the top layer in half.

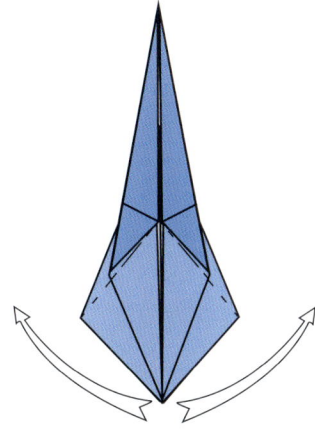

15. Inside reverse fold the wings out.

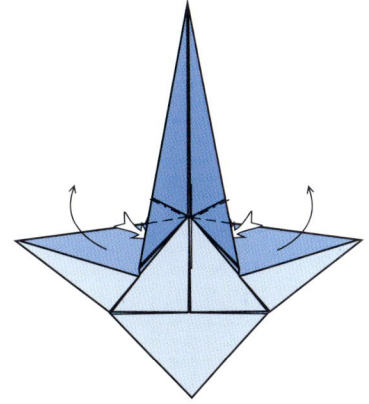

16. Valley fold the flaps at the front of the wings up while squash folding at their bases.

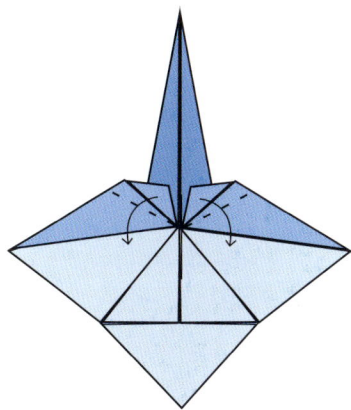

17. Valley fold the front down as far as you can.

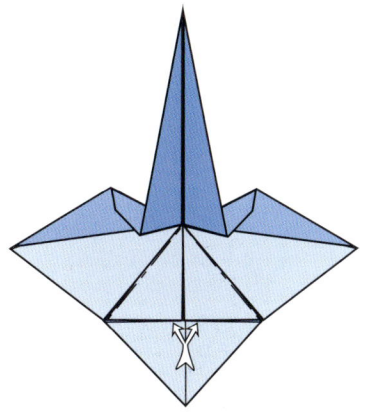

18. Reverse fold the two flaps out along the inside lines of the wings.

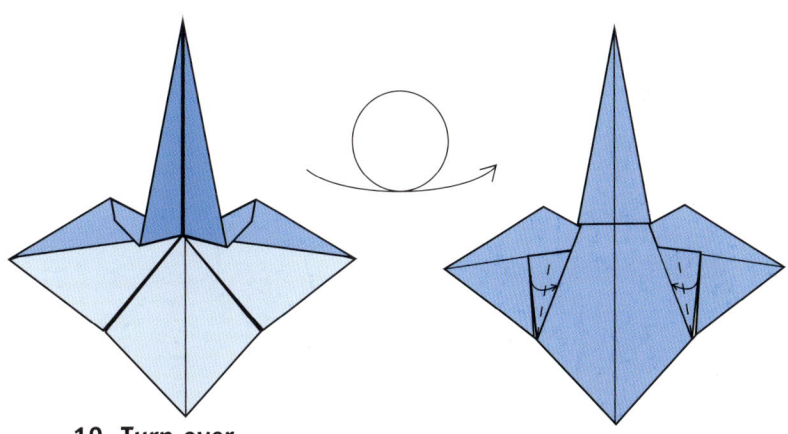

19. Turn over.

20. Valley fold the two flaps in half.

45

A Few Good Darts

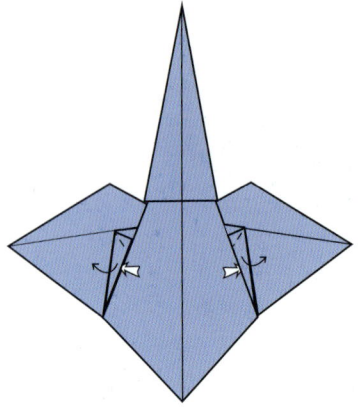

21. Open the flaps and squash fold them.

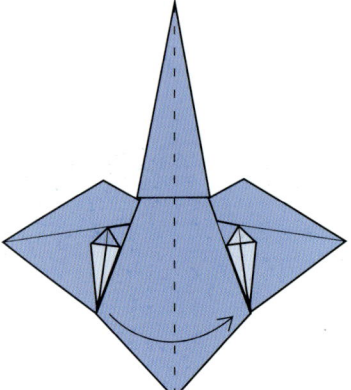

22. Valley fold the airplane in half.

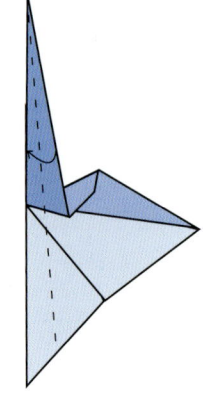

23. Valley fold the wings down by bisecting the front.

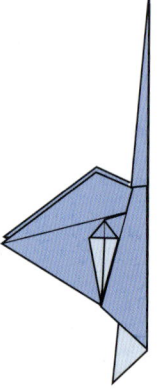

24. Open out the wings and prop up the small stabilizers—you're done with the Bird Base Attack Aircraft.

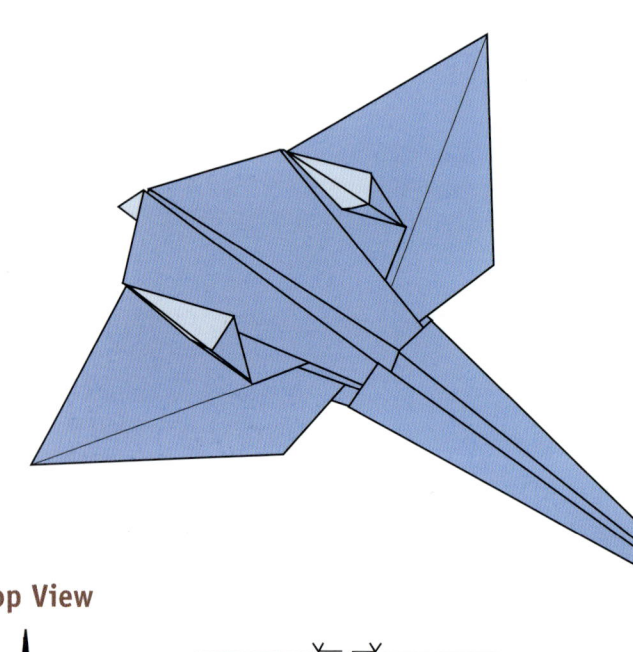

Top View

Front View

Side View

Caution: Cape does not enable user to fly.
— *Warning label on a Batman costume*

46

Stinger

I usually keep prototypes of new airplanes and try to piece them together later. This one was difficult, as the landmark for positioning the wings is subtle, and took a while to figure out. Worse yet, it was the only one that, in this situation, made a flyable airplane.

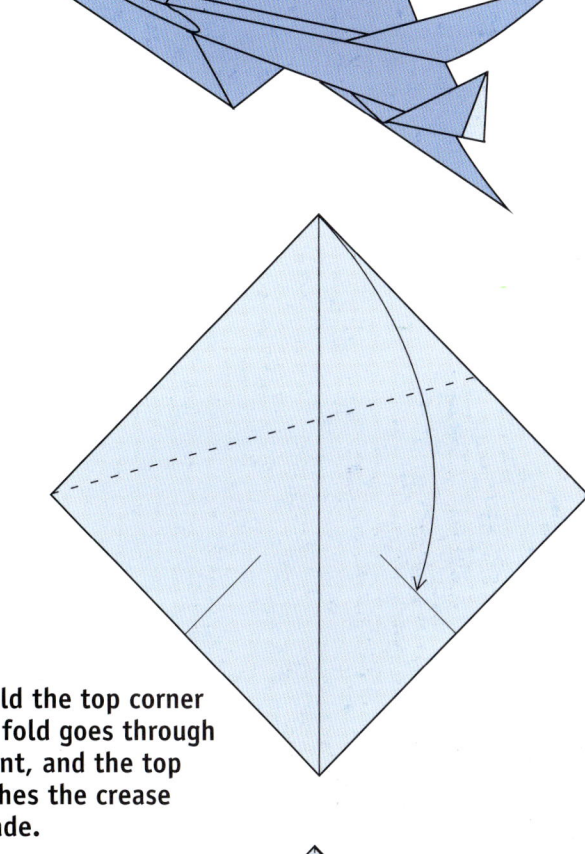

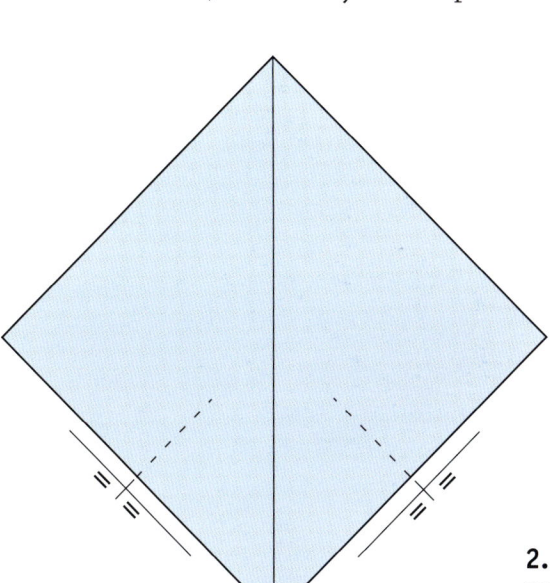

1. Valley fold each side partway, and as lightly as you can manage.

2. Valley fold the top corner so that the fold goes through the left point, and the top corner touches the crease you just made.

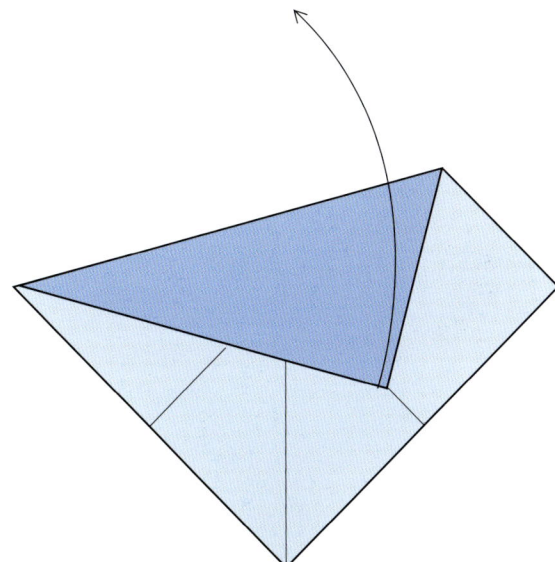

3. Unfold and repeat on the other side.

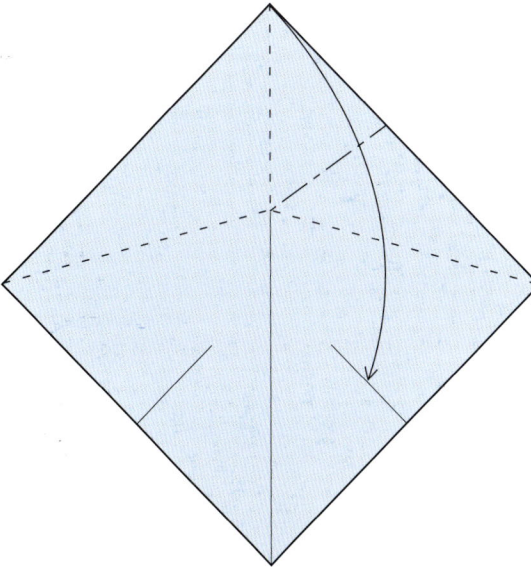

4. Rabbit ear the top along the creases you made previously.

A Few Good Darts

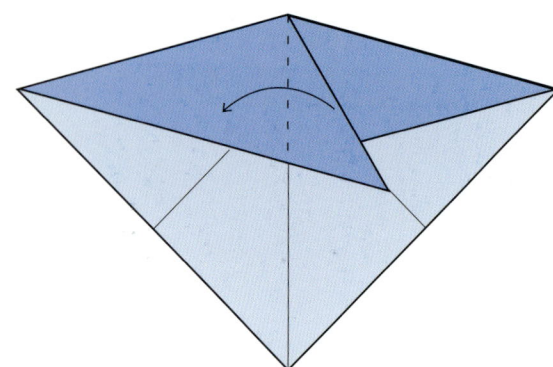

5. Valley fold the rabbit ear across the center.

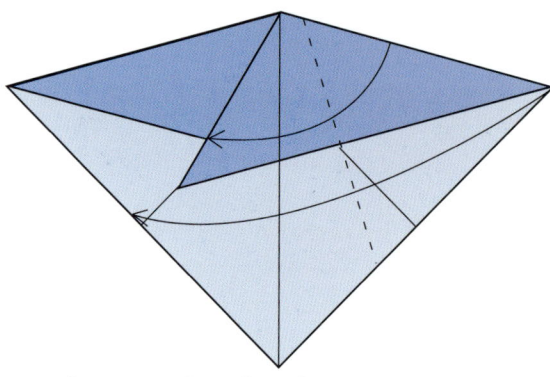

6. Okay, now it's time for the tricky landmark. Valley fold so that the lateral point touches the opposite edge. Line up the top of the flap so that it lies on the intersection shown.

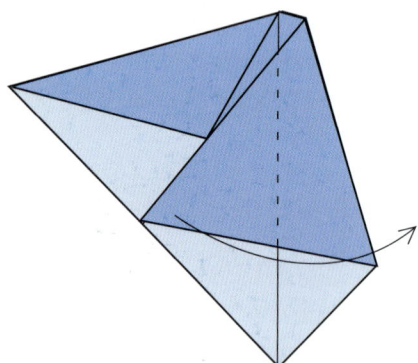

7. Valley fold the flap back along the centerline.

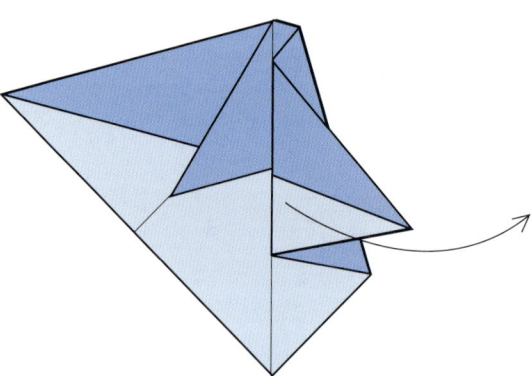

8. Unfold to step 6.

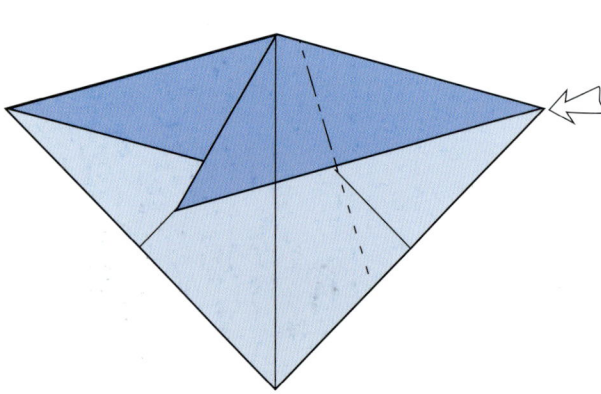

9. Reverse fold the flap in along the crease made in step 6.

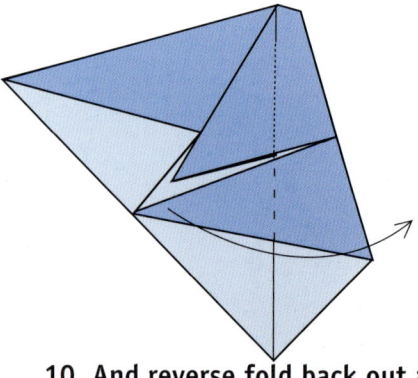

10. And reverse fold back out along the crease made in step 7.

48

Stinger

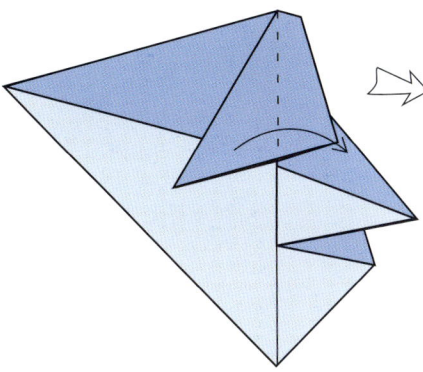

11. Valley fold the uppermost flap to the right.

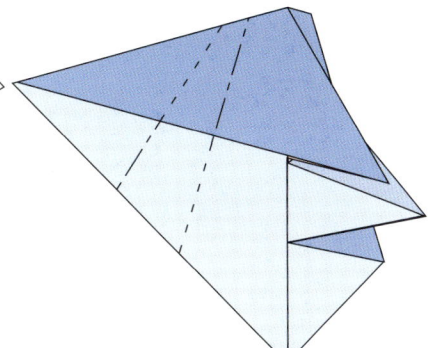

12. Repeat steps 6–11 on the other side.

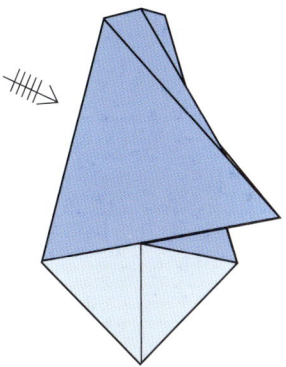

13. Repeat steps 7–11 on the other side.

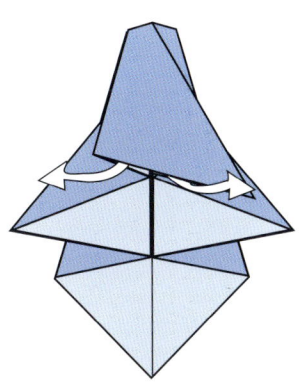

14. Undo and flatten out the top.

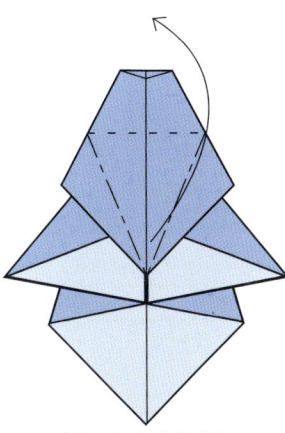

15. Petal fold.

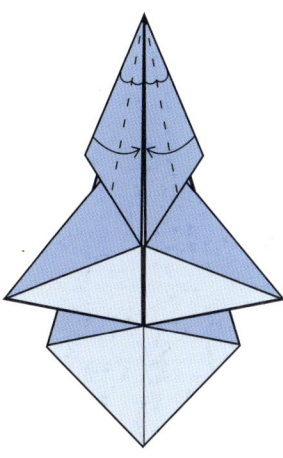

16. Valley fold each side of the top section in half. You'll make a pair of small squash folds in the process.

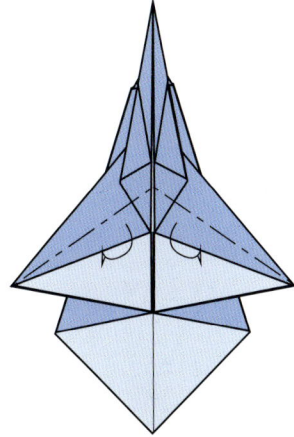

17. Mountain fold the indicated flaps behind. In addition to thickening the forward edges of the wings, this will move weight forward.

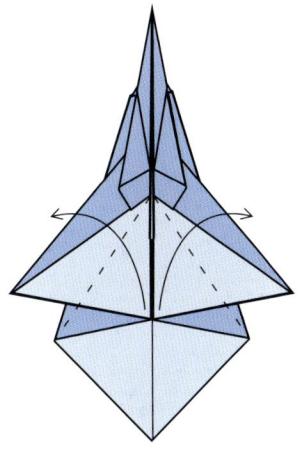

18. Valley fold the flaps.

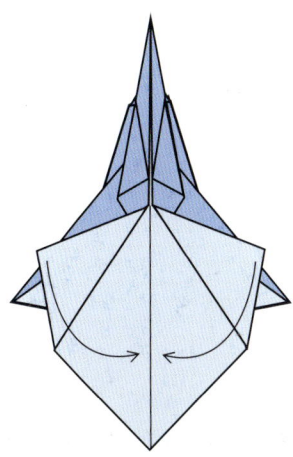

19. Unfold.

49

A Few Good Darts

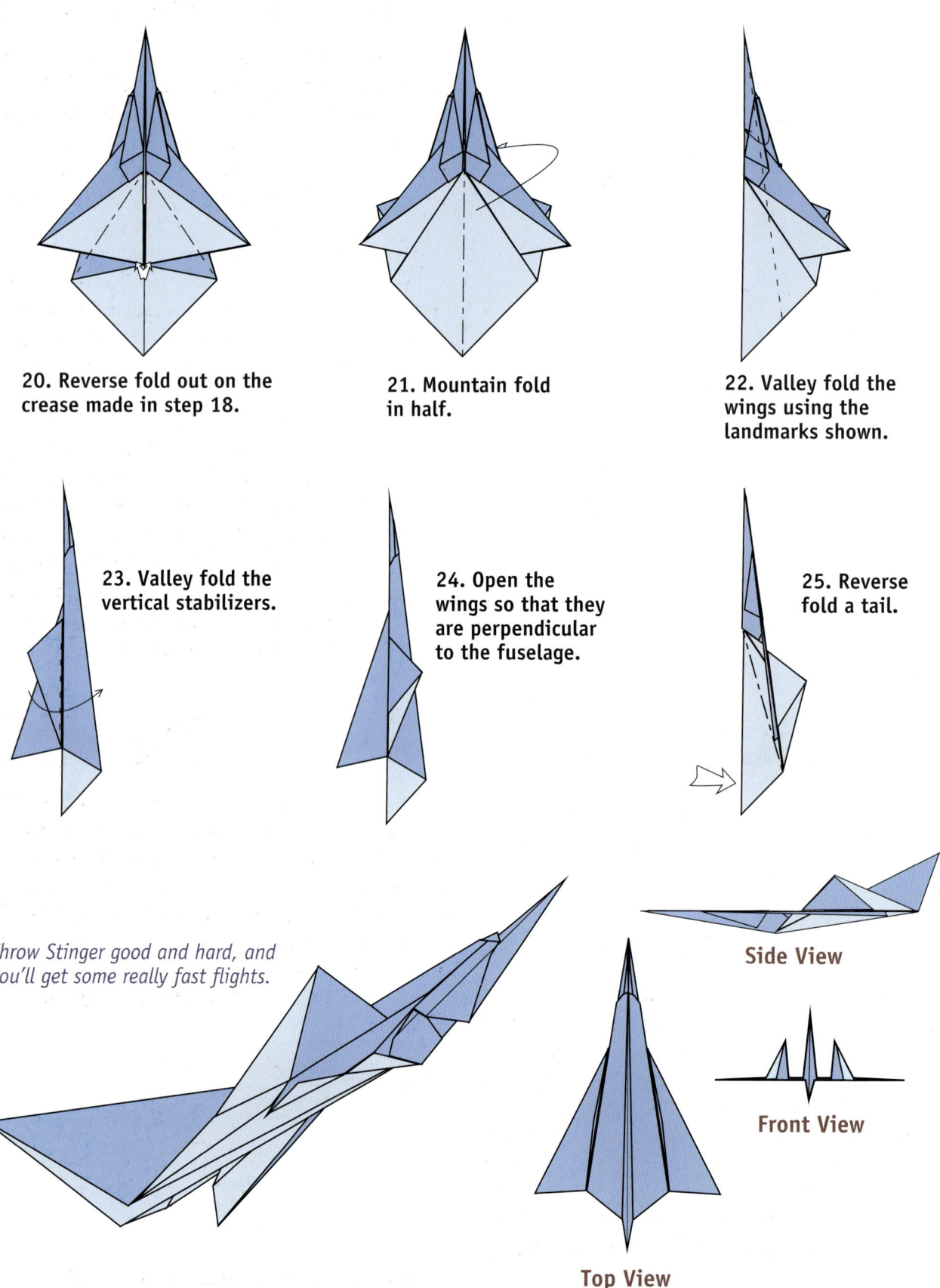

20. Reverse fold out on the crease made in step 18.

21. Mountain fold in half.

22. Valley fold the wings using the landmarks shown.

23. Valley fold the vertical stabilizers.

24. Open the wings so that they are perpendicular to the fuselage.

25. Reverse fold a tail.

Throw Stinger good and hard, and you'll get some really fast flights.

Side View

Front View

Top View

Going Fast

This section features lots of sleek darts that fly fast. But what makes a real airplane go fast? The answer is actually quite a bit more complex than you might think.

Engine

You might say that putting a more powerful engine on an airplane is what one should do to make it go faster. Indeed there is no substitute for more power. On the other hand, there are paper airplanes in this book that will do no better thrown hard. The same is true for real airplanes. For example, there are Piper Cherokees with much more powerful engines than mine; however, they only fly incrementally faster. What the larger engines give them is the ability to lift more weight.

Larger power plants can give designers a real headache too. More powerful engines are thirstier, and need more gas to stay aloft. More gas means more weight, cutting into the payload the aircraft can carry. Most airplane designs are a compromise between power, gas, and payload. Pilots of go-fast airplanes often have to choose between fuel and payload—they can't have both.

Landing Gear

The best way to make an airplane go fast is to make it slippery. The Gee Bee pictured was state of the art in the 1930s and shows some of the aerodynamic refinements that made airplanes as fast as they are now. Notice the spats covering the landing gear. Wheels, hubs, and brakes are draggy in nature. The wheel covers smooth out their profile and make them more slippery. Many airplanes use these; many more simply retract the landing gear into the wings or fuselage.

Wings

The thin stubby wings are another thing to notice. Smaller, thinner wings mean less lift, and thus less drag (lift makes drag). It just means you have to fly that much faster to stay aloft. Many aircraft use this principle to go faster. The F104 Starfighter was basically a giant engine with stubby little wings, and could fly like greased lightning. Of course, that also meant that it had to land fast, and landing accidents killed numerous unlucky pilots. Many experimental GA aircraft act similarly. Experimental simply means they don't fit the FAA's stringent rules for certification, which include the ability to land at a specified velocity. Many experimental aircraft use thin wings to go fast, but must land fast as well. They can use up quite a bit of runway doing so, and are very dangerous if the airplane is forced down away from a landing strip. The amount of energy that must be dissipated in a crash increases with the square of the velocity. That said, most crashes of experimental aircraft are caused by the same factor that causes certificated aircraft to crash—pilot mistakes.

Aerodynamics

A giant radial engine is not the most aerodynamic thing in the world, but it was what they had in the day of the Gee Bee. Notice the speed ring around it? They had to deal with the big, round engine, but made it as aerodynamic as they could. Heck, the whole airplane was a fuselage with a couple of wings, as slippery as they could make it. No, struts and wires were also not very aerodynamic,

The Gee Bee, a racing airplane from the 1930s.

but that's how it was back then. Airplanes now have all manner of covers and fairings to reduce the drag generated by the airframe. Sharp edges and things sticking out in the air stream are avoided as much as possible. In some aircraft the step to climb in retracts in flight.

Materials

Airplanes started out being made with wooden frames and fabric coverings, and many are still made the same way. These can make lots of edges and whatnot, and aren't the most aerodynamic. Aircraft made from wood could actually be far more aerodynamic, as a skilled carpenter can give a wooden surface a smooth varnished finish. The Mooney Aircraft Company has been producing single-engine retractable-gear airplanes for decades. They started out with wooden wings, but converted to metal when the highly skilled carpenters needed for their repair became scarce. The move to metal wings cost the aircraft a good 10 percent of its speed.

Many aircraft are made of aluminum due to its light weight. Pieces of metal have to be attached to each other to build an aircraft, though, which means they must be welded or riveted. Technology to weld sheets of aluminum has only recently been developed, thus most aluminum airframes have been riveted. This can reduce the aerodynamic properties of the airframe because the rivets can wind up sticking out in the air stream causing drag. This is the case in many small aircraft.

New materials like fiberglass composites and graphite can make airplanes considerably slicker and more aerodynamic because of their smooth glassy finishes and because it's relatively easy to mold them into highly aerodynamic shapes.

High Flying

If I'm taking my Cherokee on a long trip, I try to maintain significant altitude, usually as high as the engine and I can continue to breathe and the engine can maintain full power. The airplane has less air to cut through, and thus there's less drag and greater speed.

Flying way up where the air is thin is a good way to speed up. Many aircraft fly in the stratosphere. This has two distinct advantages: most weather doesn't reach that high, and one can get huge tailwinds. But to stay that high an aircraft needs a turbo-normalized or turbine engine, which can be hard on fuel reserves (see the "Loops, Tubes, and Assorted Mayhem" section for a discussion of aircraft engines). What's more, the aircraft requires an oxygen supply for the pilot and passengers, as the air in the stratosphere is too thin for humans to breathe. Many aircraft carry an oxygen bottle and cannulae, while others (including the aircraft in which most people fly) are pressurized.

All of these factors contribute to making airplanes function as they do. Airplanes are designed for distinct purposes. Some are built to fly low and slow and land short, to be useful on unimproved strips in the backcountry. Others are built to fly high and fast, to fly from one big city to another, and require thousands of feet of asphalt runway in order to land. We pilots say that one always needs the correct aircraft for one's mission.

Never feel sorry for anyone who owns an airplane. — Tina Marie

Fractal

Fractals are wonderfully complicated mathematical algorithms that generate geometric forms with iterative properties, and are frequently used in origami. Those who are arithmetically challenged can just fold the airplane, which requires no math, is more fun than math, and flies better than the famous Mandelbrot set (unless you generate a Mandelbrot set, print it, and use it to fold a paper airplane).

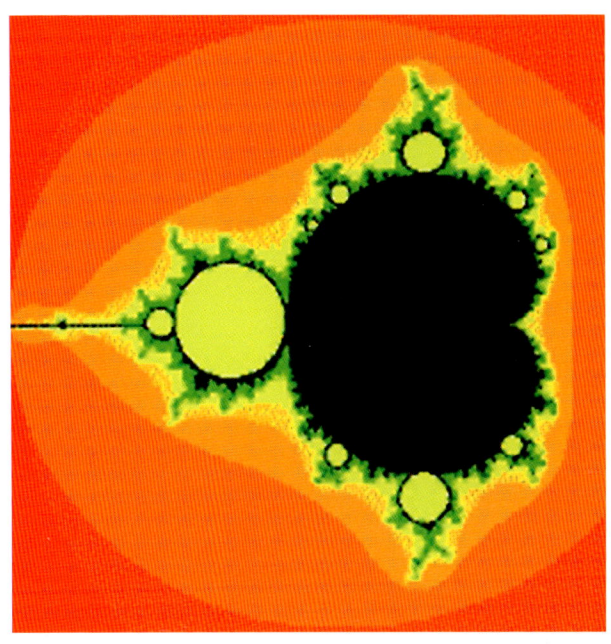

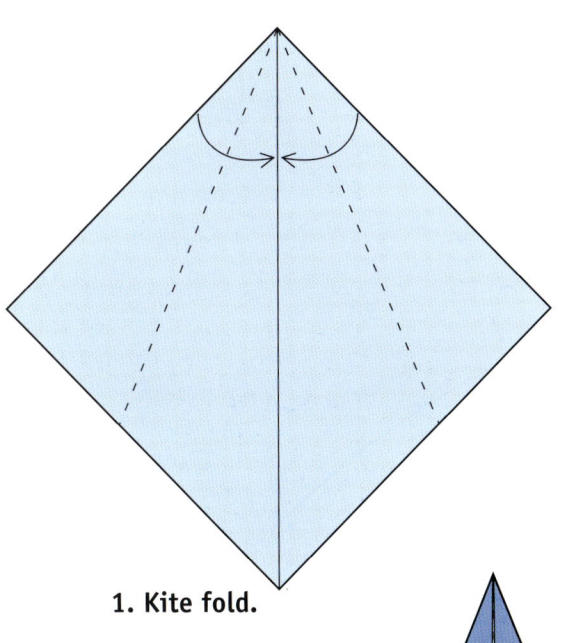

1. Kite fold.

2. Valley fold the outer folded edge to meet the inner raw edge below it. Crease and unfold. Repeat on the other side.

3. Rabbit ear along the creases you made in the previous step.

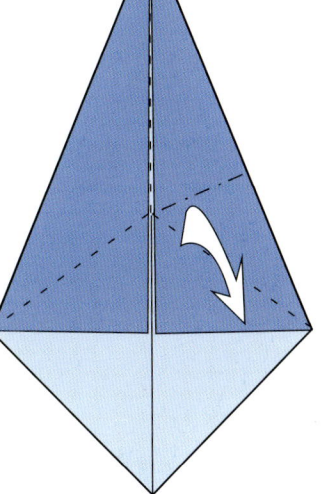

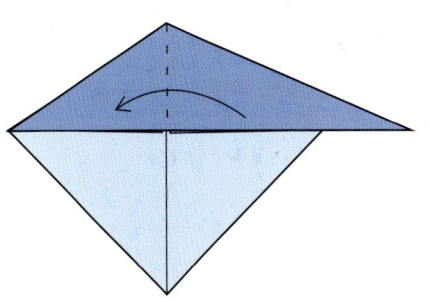

4. Valley fold the rabbit ear to the other side.

A Few Good Darts

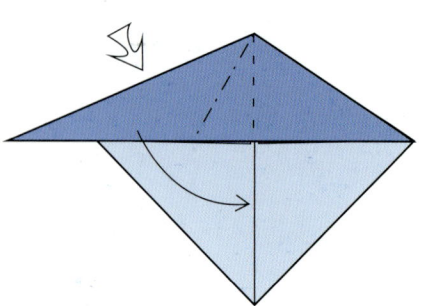

5. Squash fold the rabbit ear.

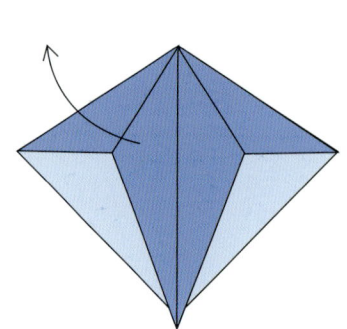

6. Unfold.

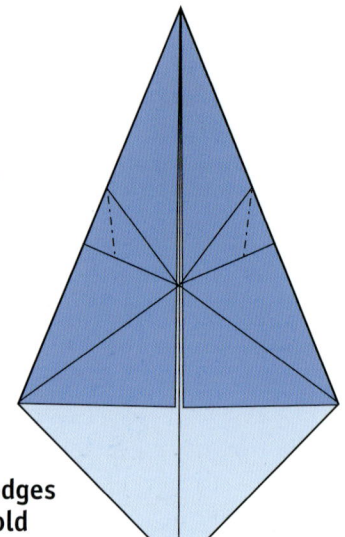

7. Mountain fold so that the edges meet the indicated creases. Fold only partway, and crease well. Then reassemble to step 6.

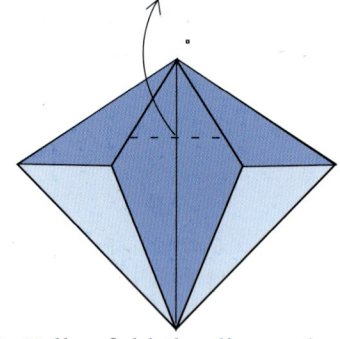

8. Valley fold the diamond upward. Pockets will form from the paper underneath.

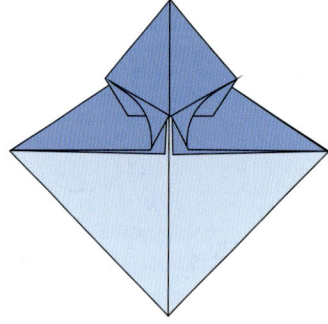

9. This shows the step in progress. The pockets can be flattened out along the creases made in step 7.

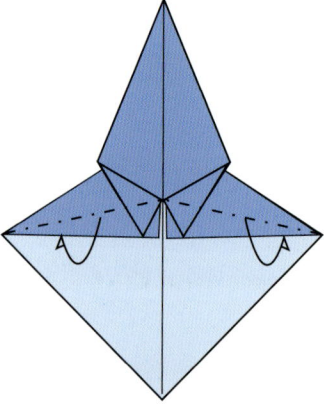

10. Mountain fold the flaps behind to move the weight of the Fractal forward.

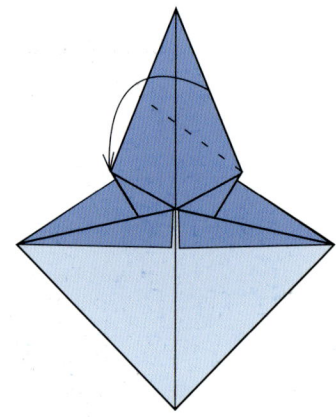

11. The first iteration is complete. Valley fold the upper flap so that its outer edge touches its corner.

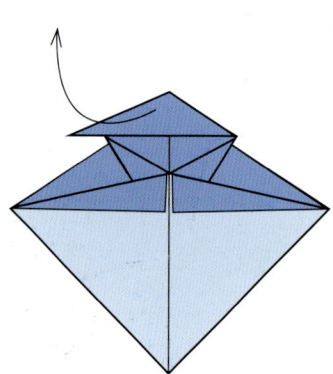

12. Unfold.

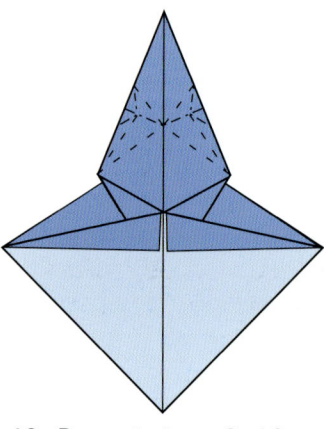

13. Repeat steps 2–10 on the upper flap to perform the second iteration.

54

Fractal

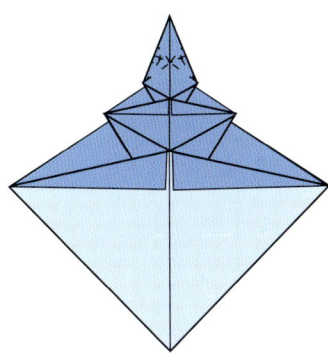

14. The second iteration is done. Yep, you guessed it—repeat steps 2–10 again.

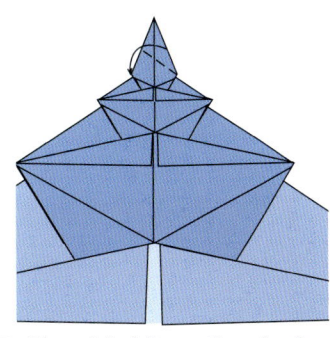

15. The third iteration is done. You now have firsthand knowledge of iterative properties. Valley fold the top as in step 11.

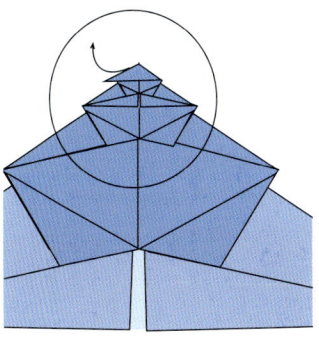

16. Unfold and repeat on the other side.

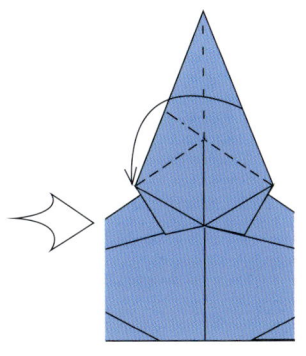

17. Rabbit ear the top.

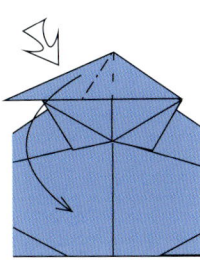

18. Squash fold.

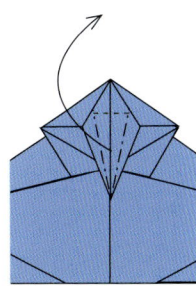

19. Petal fold the flap back up.

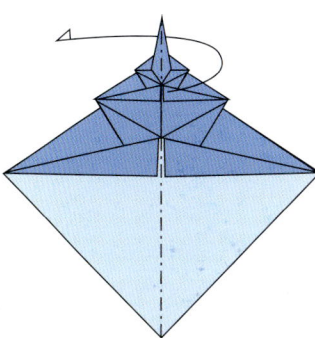

20. Mountain fold the fractal in half.

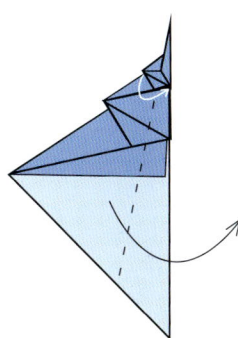

21. Valley fold the wings by bringing together the points shown on the white arrow. Do this carefully, as the front of the Fractal becomes quite thick with iterated layers.

22. Reverse fold up a tail.

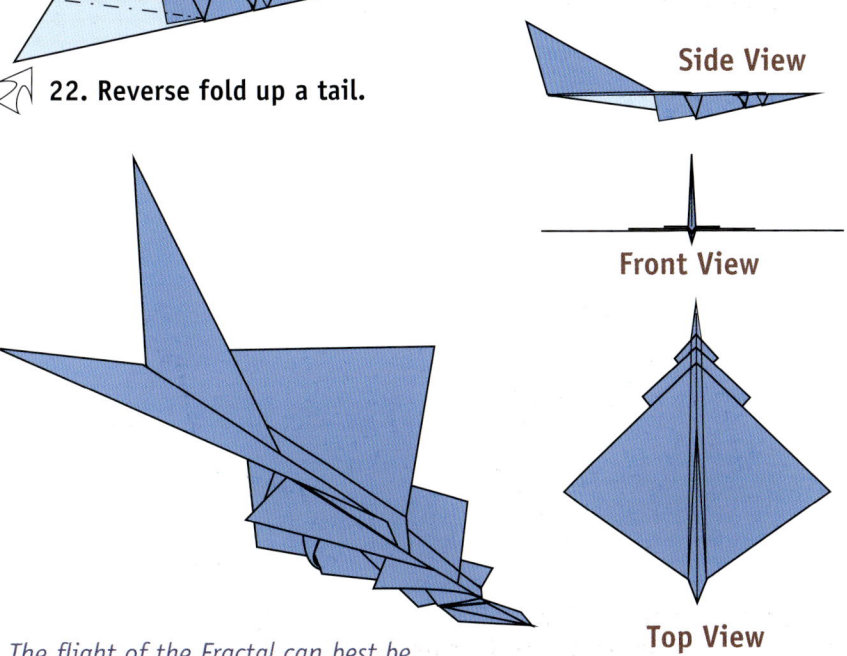

The flight of the Fractal can best be described by the equation $z := z * z + c$.

55

Canards

Elbert Leander "Burt" Rutan is one of the most visionary aircraft designers of our times. His efforts run the gamut from successful home-built aircraft to the first civilian spaceship.

Born in central California, Rutan showed an early interest in aviation, soloing an aircraft at age sixteen. He graduated third in his class at the California Polytechnic Institute, and worked as a flight test project engineer for the US Air Force.

He moved into the world of experimental aviation when he worked for Bede Aircraft in Newton, Kansas, but he soon set out on his own, establishing the Rutan Aircraft Factory and later, Scaled Composites.

Burt Rutan

Rutan quickly showed his aptitude for designing canard aircraft (Bleriot had designed the first canard (French for duck. His aircraft looked a lot like a duck. Really.). Rutan came out with designs for wood canards to be sold to experimental builders. His designs really flourished when he incorporated the reinforced fiberglass construction technology used to make high-performance sailplanes. The engine that made my Cessna 150 fly at 90 mph could push a Varieze almost 200 mph. His aircraft were revelations, and the Rutan Aircraft Factory, and later Scaled Composites, created a number of imaginative canard designs, including the Voyager, the first airplane to fly around the world without refueling, and the Global Flyer, which accomplished the same feat flown solo.

The Rutan Varieze

But nothing has captured the imagination of the public more than the space program developed by Rutan and carried out by Scaled Composites. Rutan analyzed the X-15 program pursued by the US Air Force, and developed a shuttlecock-like mechanism to bring a spacecraft in for reentry in lieu of a heat shield. Backed by Microsoft founder Paul Allen, Rutan developed a space vehicle and an aircraft to carry it high in the atmosphere, and on October 21, 2004, Burt's friend and pilot, Mike Melville, became the first private citizen to earn astronaut wings.

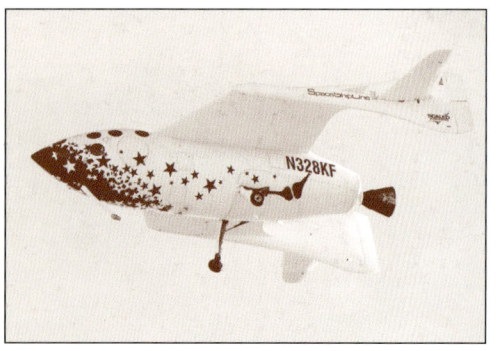

Space Ship One

Rutan has more recently teamed up with the ubiquitous billionaire adventurer Sir Richard Branson to develop Virgin Galactic, which will take anyone on a suborbital space flight (for a hefty price). He has certainly left an indelible mark on the world of aviation.

Widebody

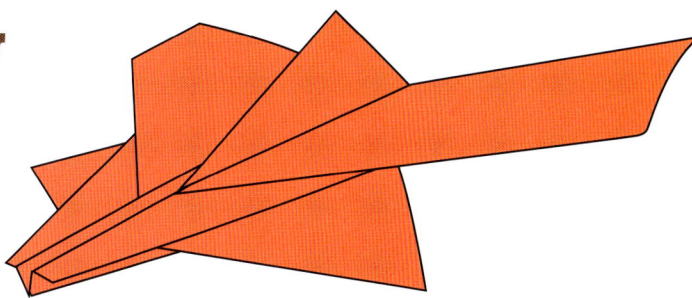

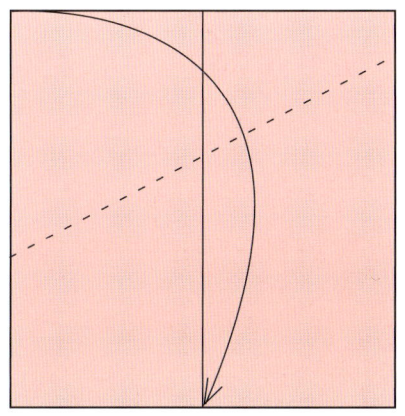

1. Valley fold the left corner to the bottom middle crease.

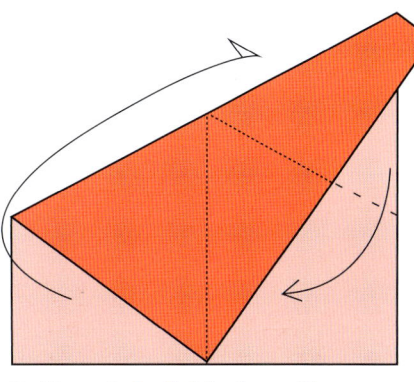

2. Mountain fold along the centerline while valley folding the right corner down to meet the left corner at the midline.

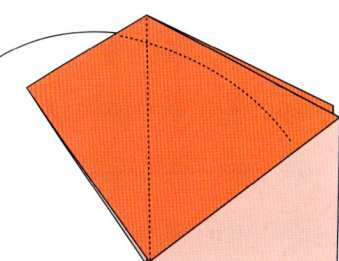

3. Mountain fold a layer out from behind.

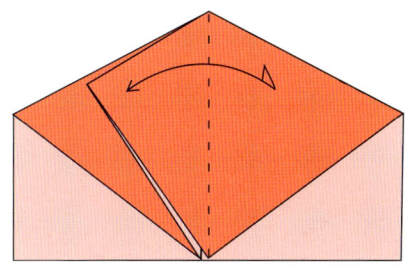

4. Valley fold the topmost flap over, crease well, and fold back.

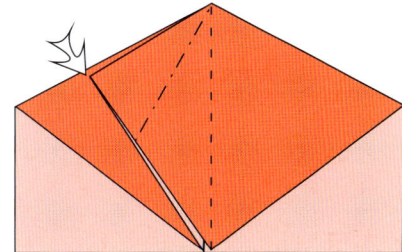

5. Squash fold.

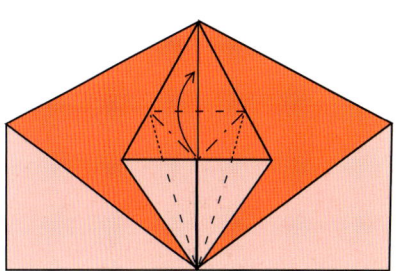

6. Petal fold.

There's no need to teach an eagle to fly. — Greek proverb

Canards

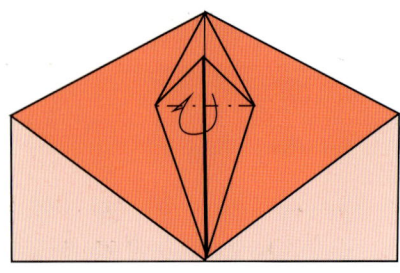

7. Tuck one layer around into the inside.

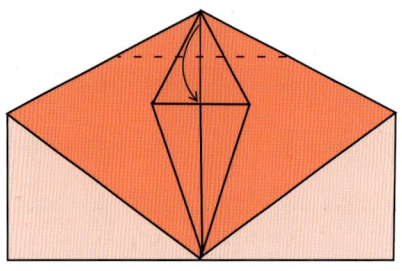

8. Valley fold so that the top meets the folded line shown.

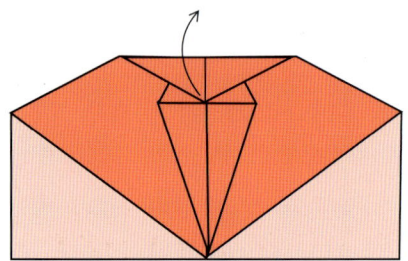

9. Crease well, and unfold.

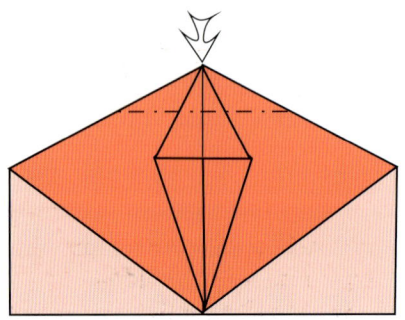

10. Open sink the top.

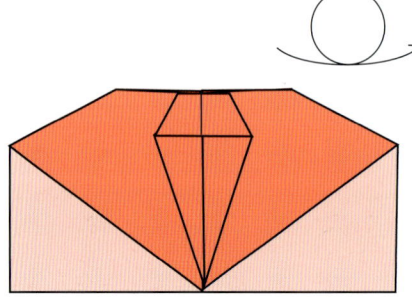

11. Turn over.

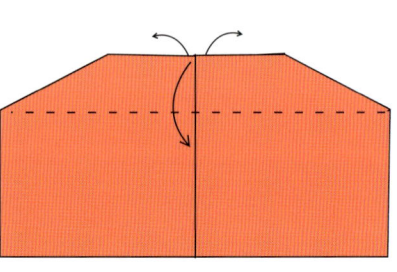

12. Valley fold the top down corner to corner, spreading out the sink fold.

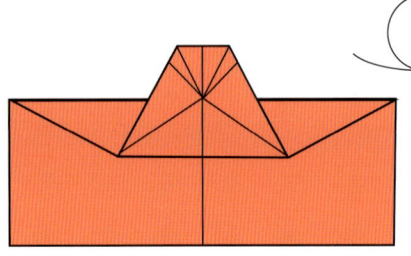

13. Turn over.

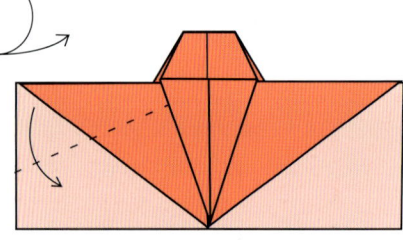

14. Valley fold following the underlying layers.

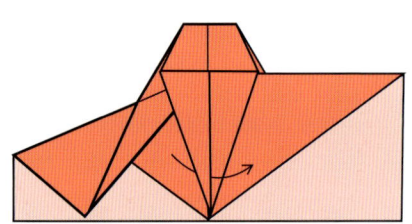

15. Fold the flap as far to the side as it will go, freeing trapped paper.

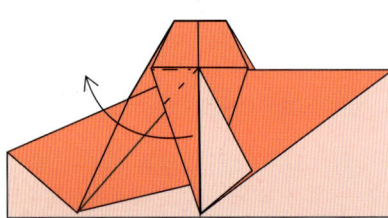

16. Swivel the flap upward so that the colored edges line up together.

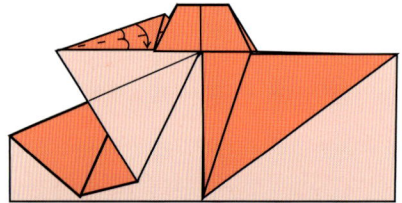

17. Valley fold the top flap in half, tucking in the middle. This will help hold the front together.

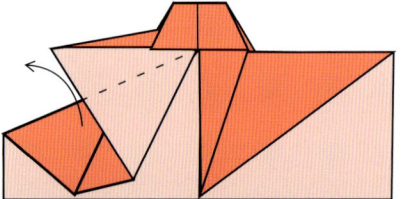

18. Valley fold the large flap upward. This should follow creases mostly already made.

Widebody

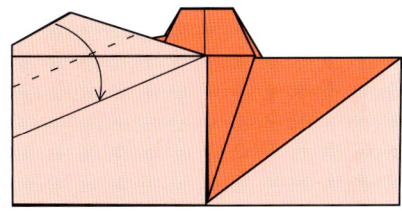

19. Valley fold the large flap in half such that the short raw edge at the top lies along the crease beneath it.

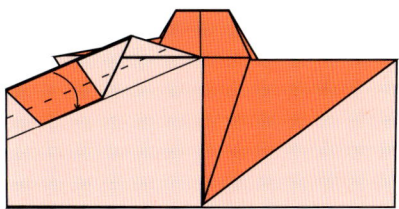

20. Valley fold in half again.

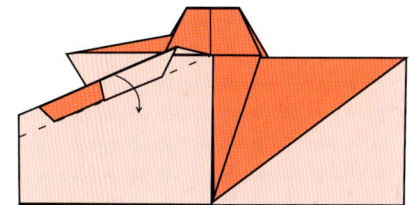

21. Valley fold back down.

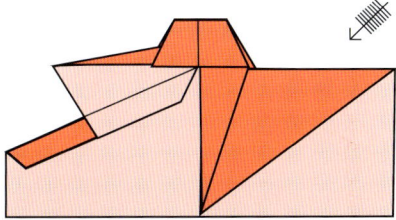

22. Repeat steps 14–21 on the other side.

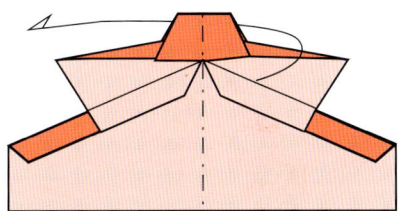

23. Mountain fold in half.

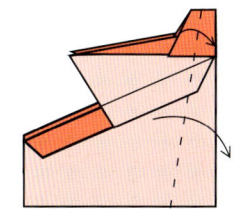

24. Valley fold the wings by lining up the folded edges in the front.

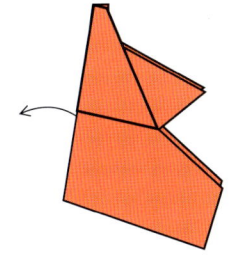

25. Unfold.

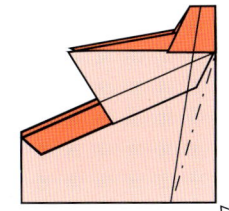

26. Reverse fold as far as you can for a tail.

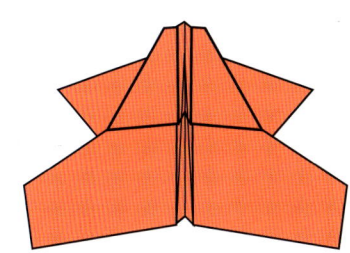

Top View

Front View

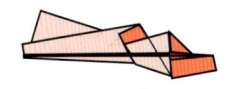

Side View

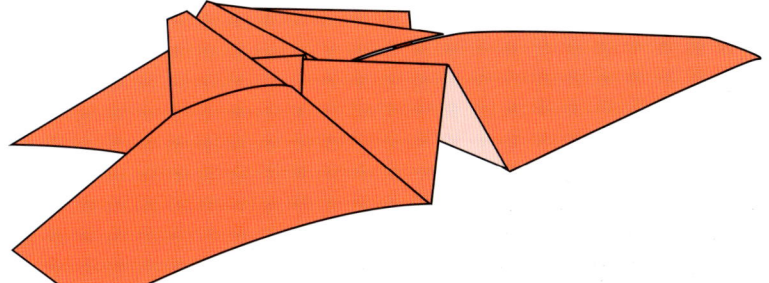

Thrown soft or hard, the Widebody will give a good flight. It glides well, and might even do some good stunts.

59

Canards

Diamondhead Staggerwing

An antique aircraft, the Beech Staggerwing is one of the most elegant ever built. This canard is similar, in that the canard wing is lower and forward of the main wing. Most real canards locate the canard wing forward and above, and are mostly used for swept-wing aircraft, to help control the airplane when flying at slow speed.

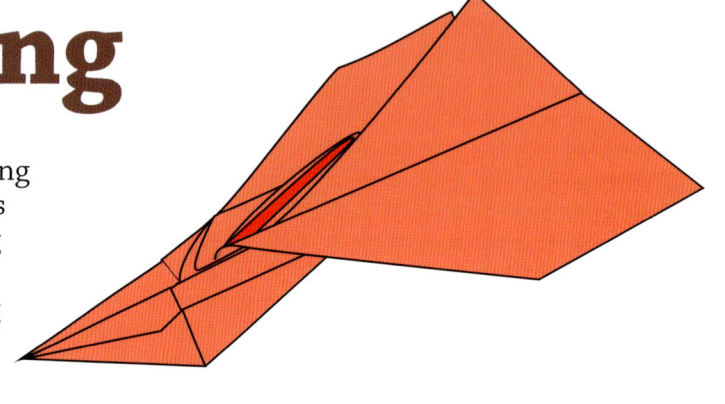

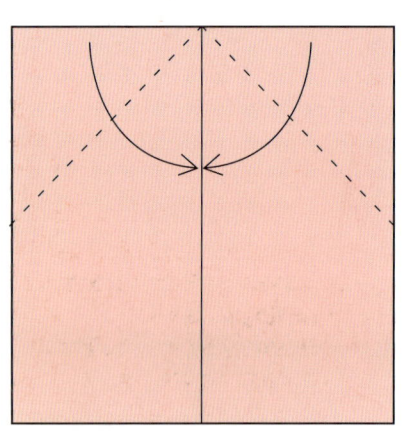

1. Valley fold both top corners into the center.

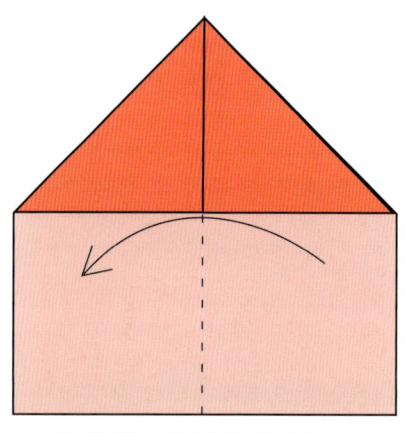

2. Valley fold in half.

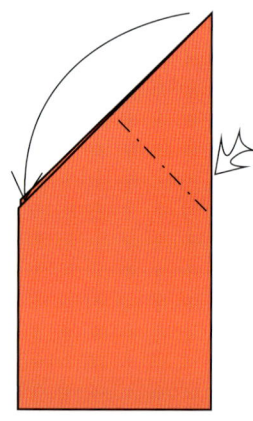

3. Reverse fold the top point down to meet the angles below it.

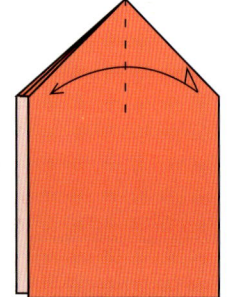

4. Valley fold partway down, crease, and unfold.

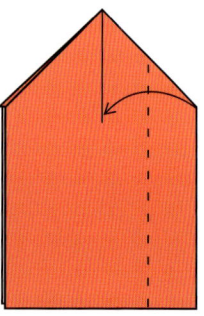

5. Valley fold the folded edge to the crease you made.

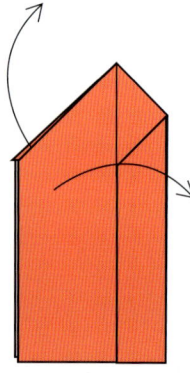

6. Open the model up, lifting the middle flap up and squash folding it.

Diamondhead Staggerwing

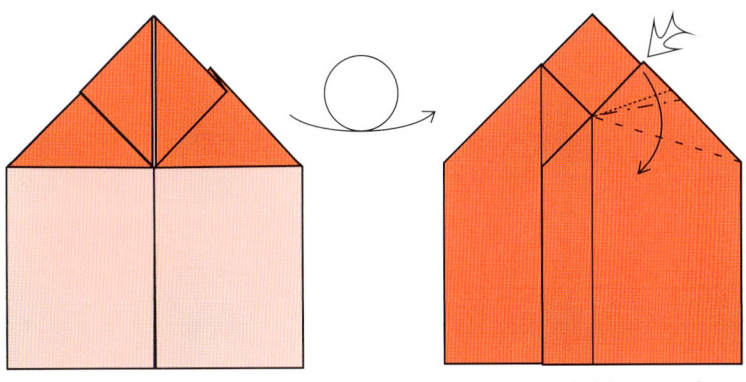
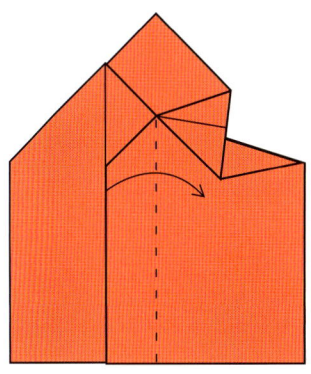

7. Turn over.

8. Squash fold one of the forward corners down.

9. Valley fold the middle flap to the other side.

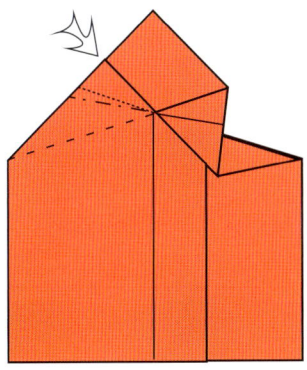
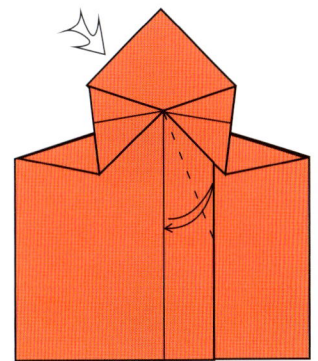
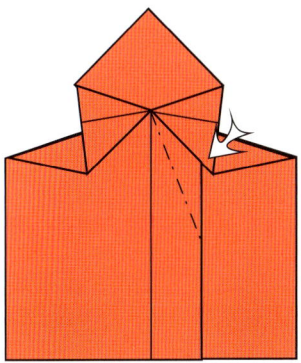

10. Repeat step 8 on the other side.

11. Valley fold the folded edge of the top flap to the centerline. Crease well, and unfold.

12. Closed sink the front of the flap along the crease you just made. This will help hold the front end together.

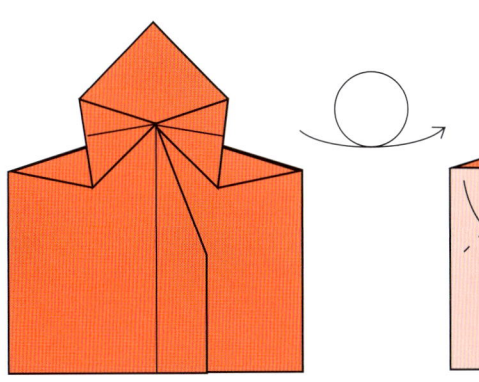
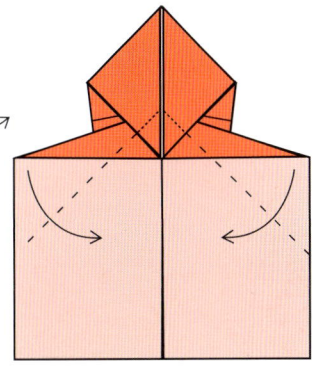
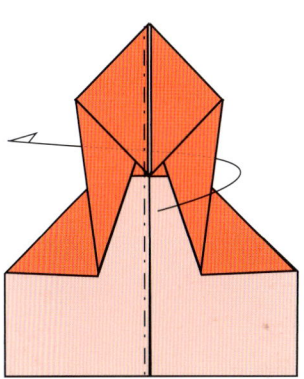

13. Turn over.

14. Fold flaps down along the folded edges of the flaps underlying them.

15. Mountain fold in half.

61

Canards

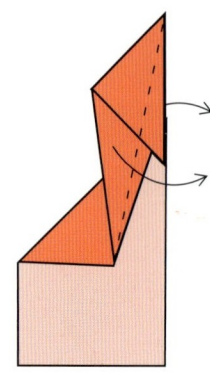 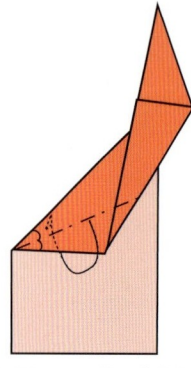 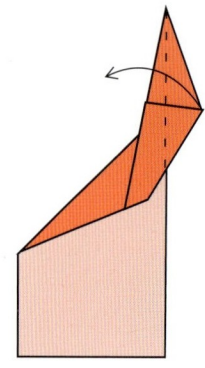 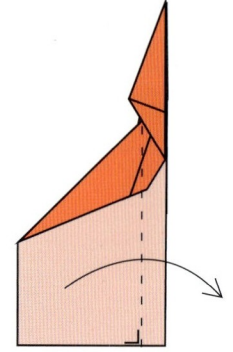

16. Valley fold a flap down in front as far as it will go. Repeat behind.

17. Mountain fold the flaps in half to concentrate the front of the main wings.

18. Valley fold the forward flaps up along the bottom edge. Again, repeat behind.

19. Valley fold the wings perpendicular to the raw edge in the back.

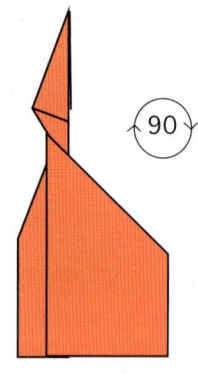 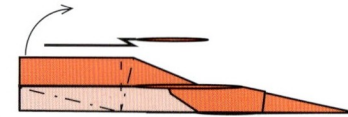 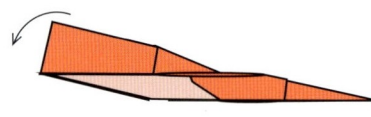

20. Pick both sets of wings up, leave them perpendicular to the body, and rotate the aircraft 90 degrees. Its time to start the difficult task of making the tail.

21. Crimp fold from the angle at the top of the tail (paper will come up from the bottom), and make certain the resulting mountain folds hit the back of the wings where they meet the fuselage. Crease well.

22. Unfold.

23. Refold using reverse folds, to lock the tail into place.

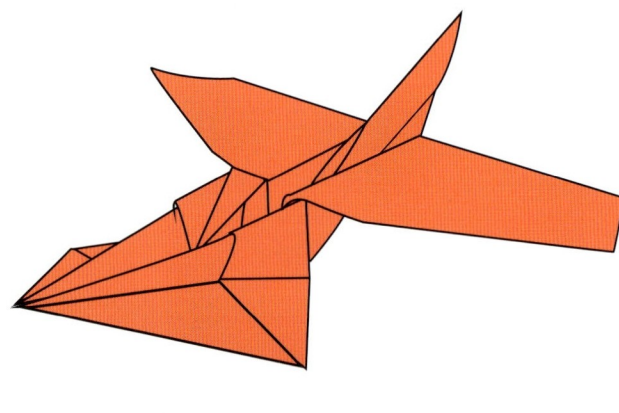

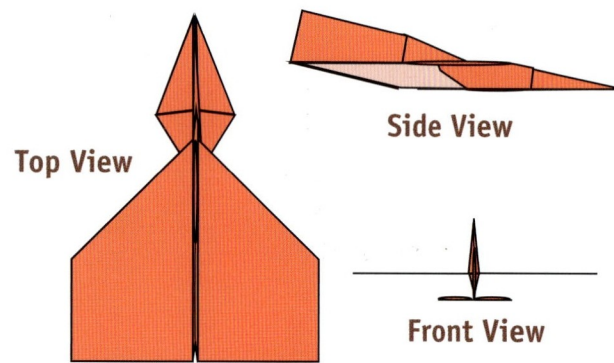

Top View

Side View

Front View

Canard with Pitot

A good canard with a little something extra.

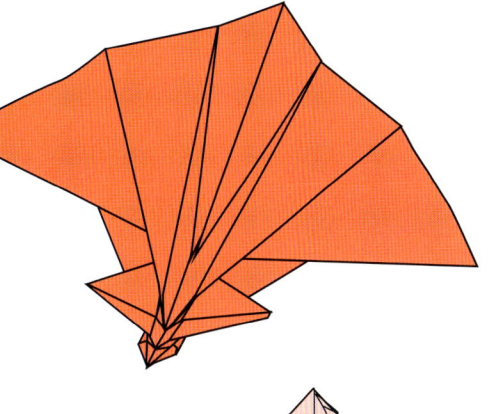

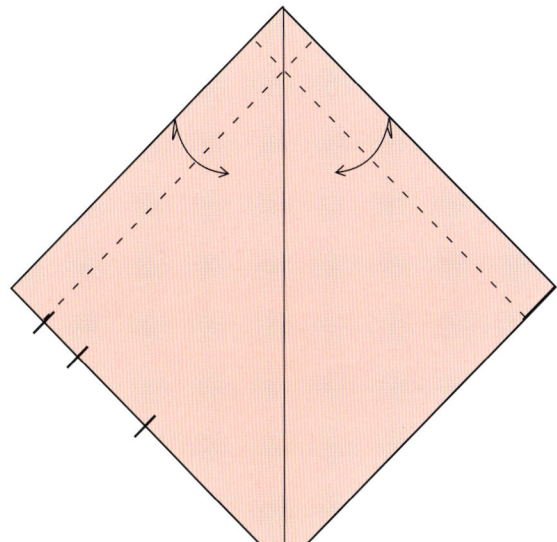

1. Valley fold $1/8$ along each edge, and unfold.

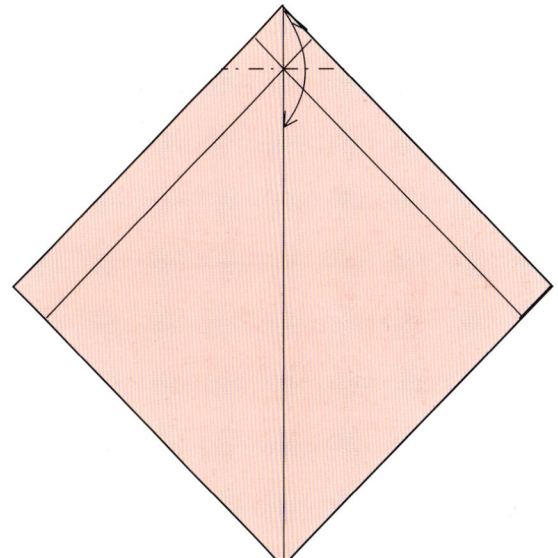

2. Mountain fold along the intersection of the three creases, and unfold.

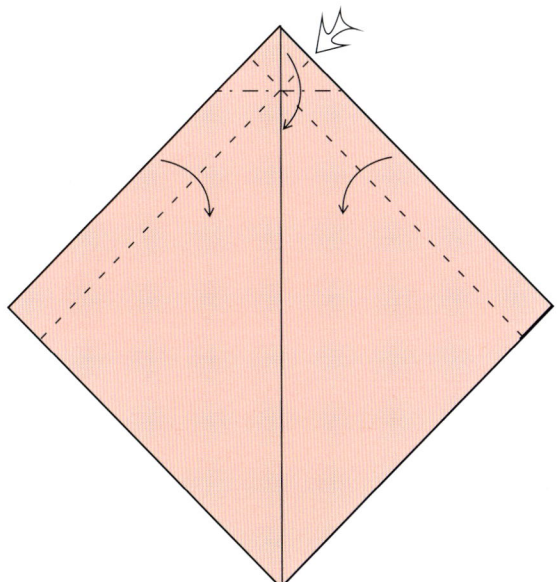

3. Fold a preliminary base at the top.

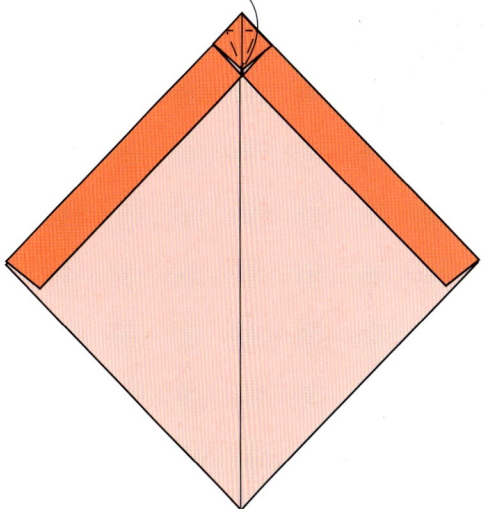

4. Petal fold upward.

Canards

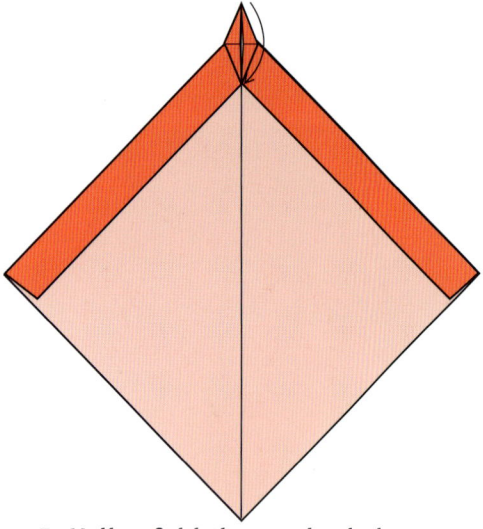

5. Valley fold the top back down.

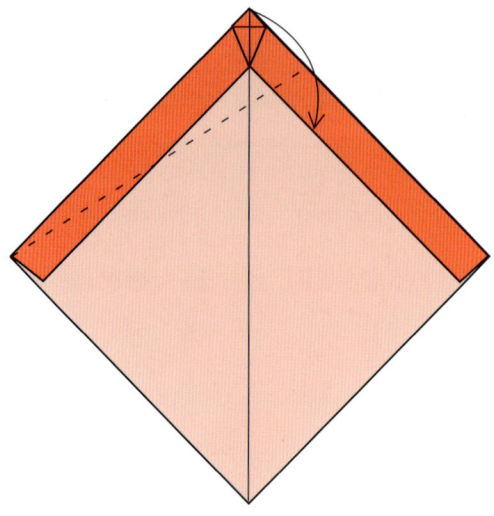

6. Valley fold so that the top point hits the colored raw edge.

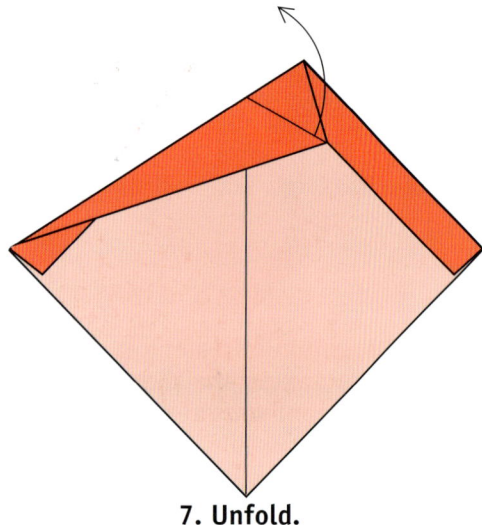

7. Unfold.

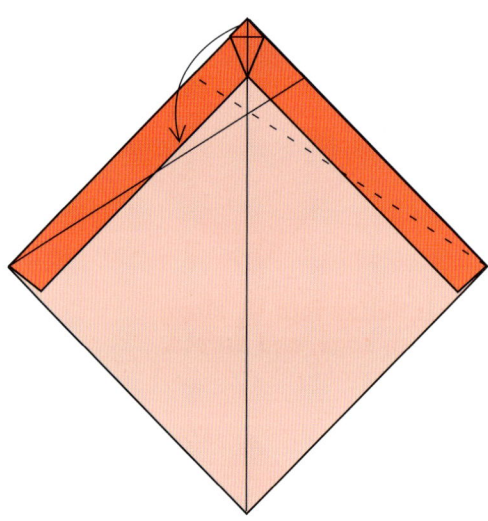

8. Repeat on the other side.

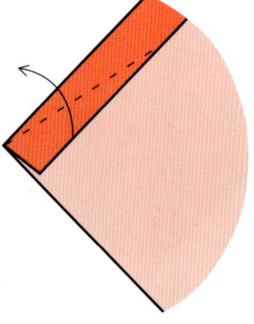

9. Detail of the left wing tip. Valley fold the colored flap up along the crease.

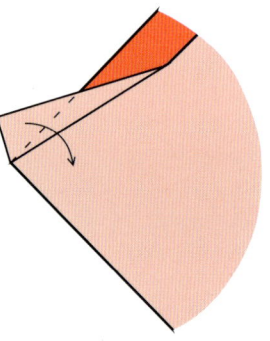

10. Valley fold back down along the folded edge beneath.

Canard with Pitot

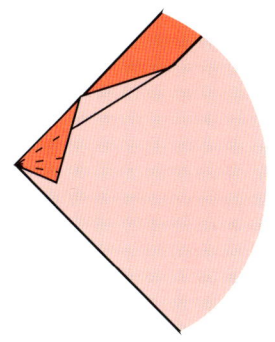

11. Continue folding back and forth three more times.

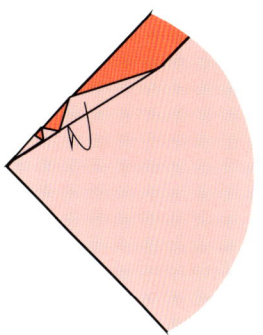

12. Then mountain fold the works behind.

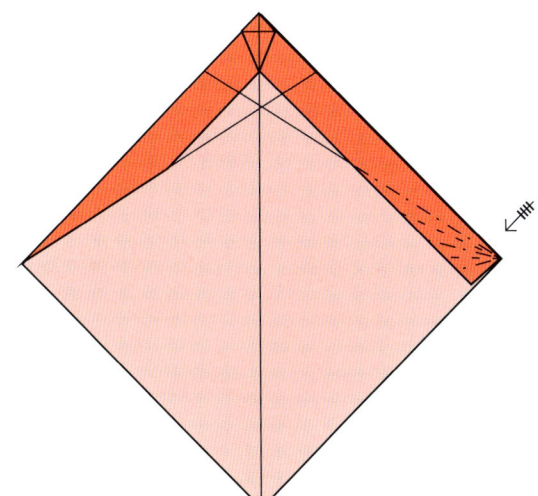

13. Repeat steps 9–12 on the other side.

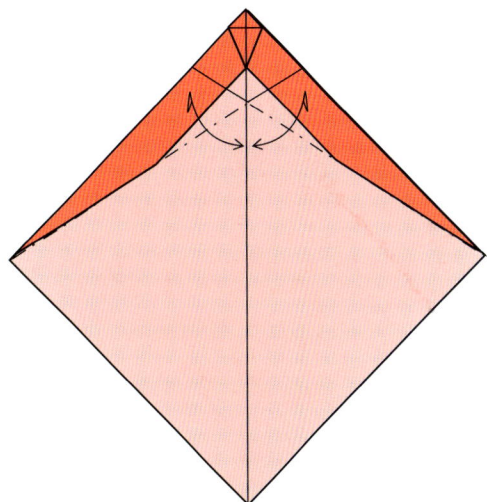

14. Mountain fold along the creases previously made, and unfold.

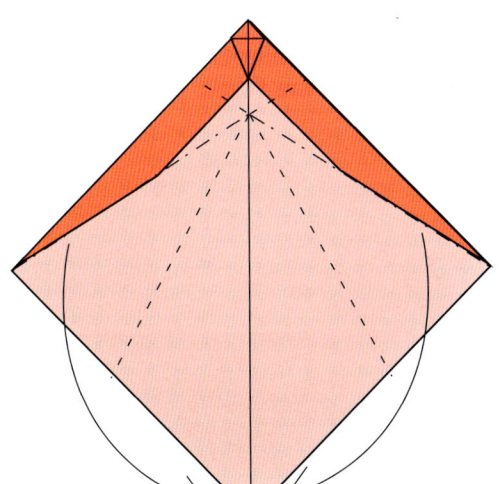

15. Fold an offset preliminary base using the creases made.

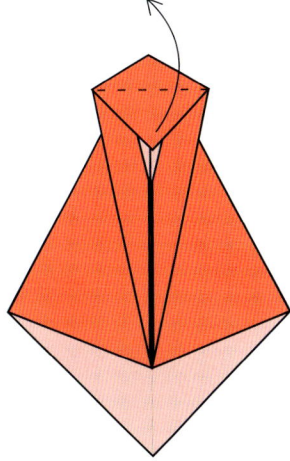

16. Valley fold the top flap upward.

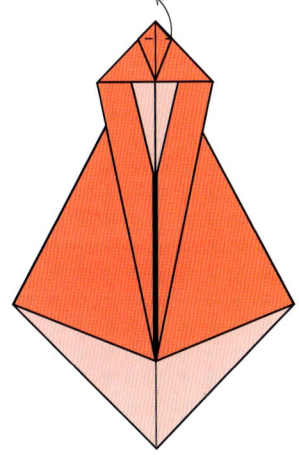

17. Valley fold the top flap upward.

65

Canards

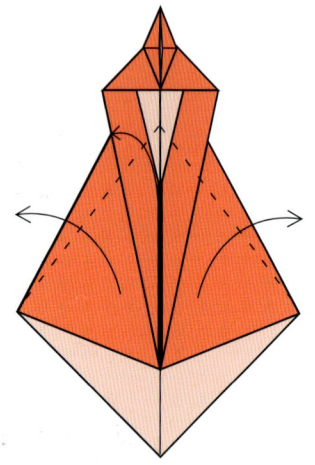

18. Valley fold the wings so that their forward edge hits the intersection shown.

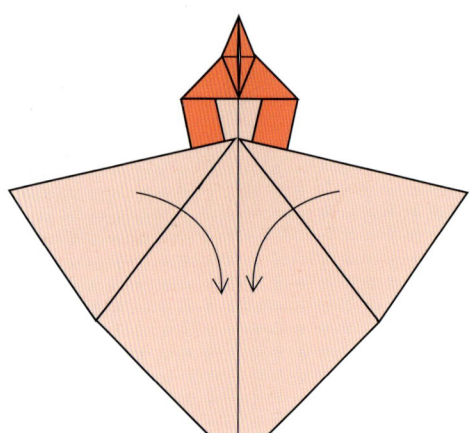

19. Fold back.

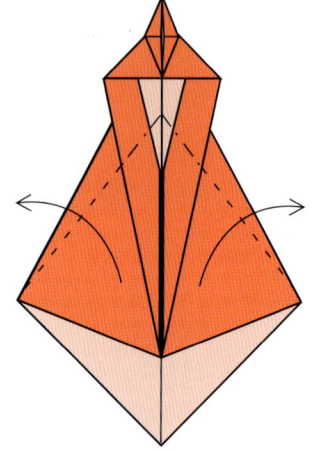

20. Reverse fold the wings out along the creases made in step 18.

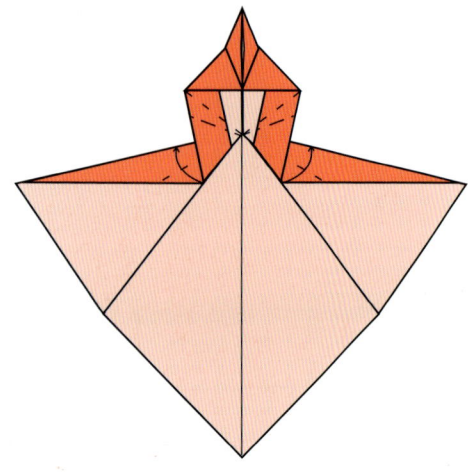

21. Swivel some paper upward.

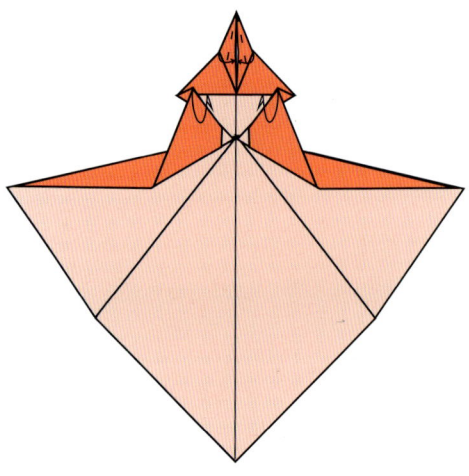

22. Tuck the upper flaps under the layer above them, and narrow the front.

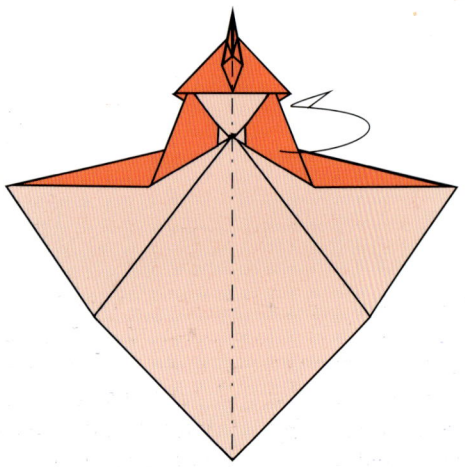

23. Mountain fold the model in half.

If God wanted us to fly, He would have given us tickets. — Mel Brooks

Canard with Pitot

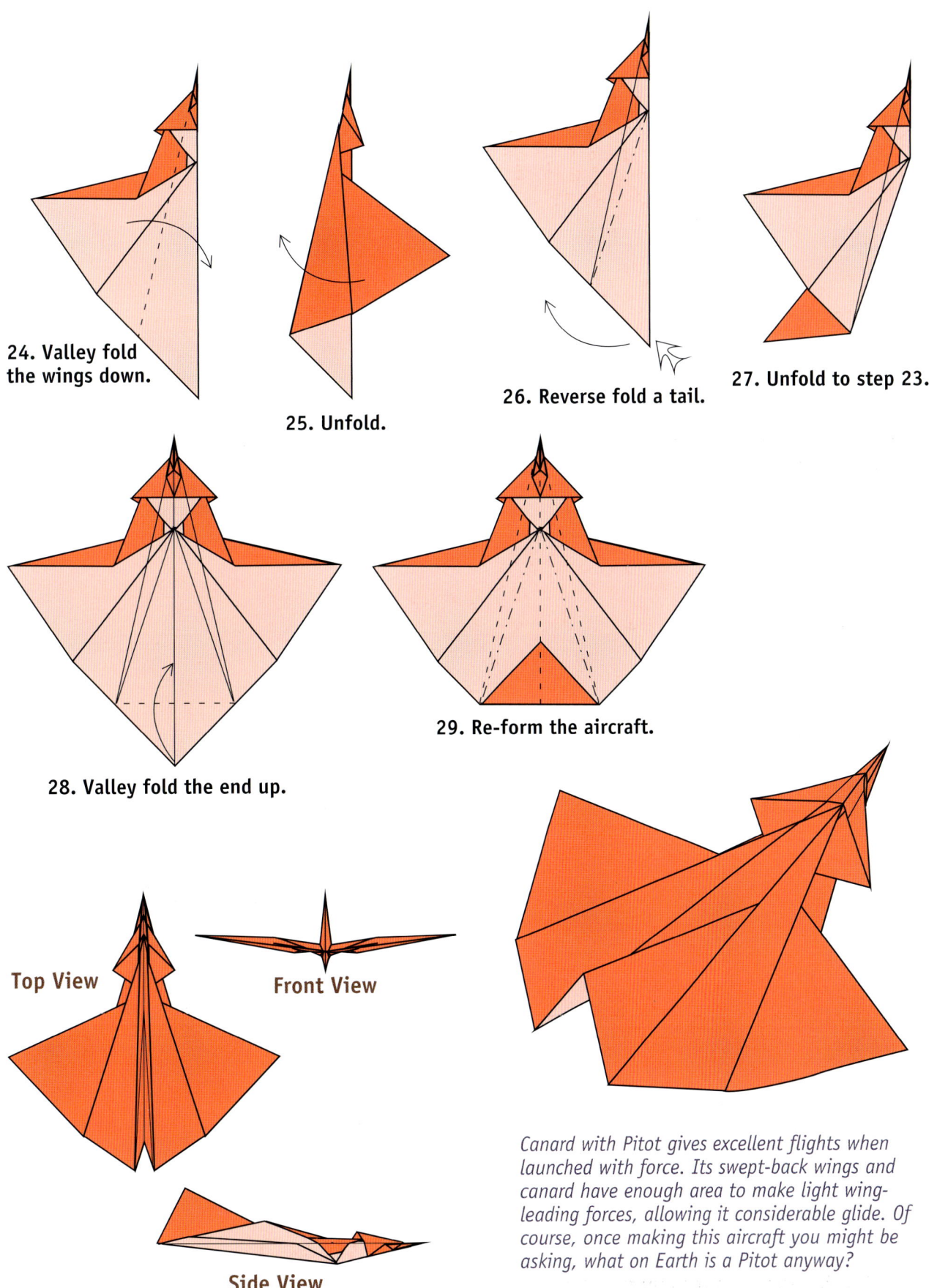

24. Valley fold the wings down.

25. Unfold.

26. Reverse fold a tail.

27. Unfold to step 23.

28. Valley fold the end up.

29. Re-form the aircraft.

Top View

Front View

Side View

Canard with Pitot gives excellent flights when launched with force. Its swept-back wings and canard have enough area to make light wing-leading forces, allowing it considerable glide. Of course, once making this aircraft you might be asking, what on Earth is a Pitot anyway?

Canards

French hydraulic scientist Henri Pitot, who was the first to measure the speed of the River Seine, invented the Pitot tube. It is basically a sphygmomanometer (pressure gauge) that is put on most airplanes. Like our canard here, the Pitot tube is placed on the nose of the F14 Tomcat, formerly used by the US Navy, and on some other fighter jets. In most light aircraft it's on the wing. The Pitot tube is placed so that the air passing the airplane (called ram air) hits it. The pressure from the Pitot tube is compared to the ambient air pressure, which is measured from the fuselage (or even the cockpit if the aircraft is not pressurized). As the aircraft speeds up, the pressure in the tube rises relative to the ambient pressure, which is read as airspeed.

Air moving quickly past the aircraft (ram air) hits the Pitot tube and causes pressure.

This system works well for most aircraft, though not for aircraft flying too slowly, as the pressure in the pitot decreases below its threshold for measurement. Velocities approaching the speed of sound are also difficult to measure with a Pitot-static system (the combination of a Pitot tube and static air measurement) because transonic speeds cause shock waves that distort the result. Altitude is also measured by pressure. Altimeters are simply pressure meters that the pilot calibrates for ambient pressure.

Just because you know your airspeed does not mean you know how fast you're actually going. Due to the wind, your speed over the ground can be quite different from your airspeed. Let's say I was flying the *Free Bird,* my Piper Cherokee, north at 120 mph. Often the winds come off Lake Erie (north of my home in Columbus, Ohio—Go Bucks!) and blow south at 20 mph. If I'm doing 120, but have a wind on my nose of 20, the actual speed of my airplane over the ground is only 100 mph, as the wind slows it down. The converse is true when I turn around to fly home; the Free Bird will get a boost from the tailwind and travel at 140 mph. I've had trips where I was going 150 mph on the outbound leg and was passed by trucks on the way back.

This is compared to the atmospheric pressure, which changes with altitude and weather conditions.

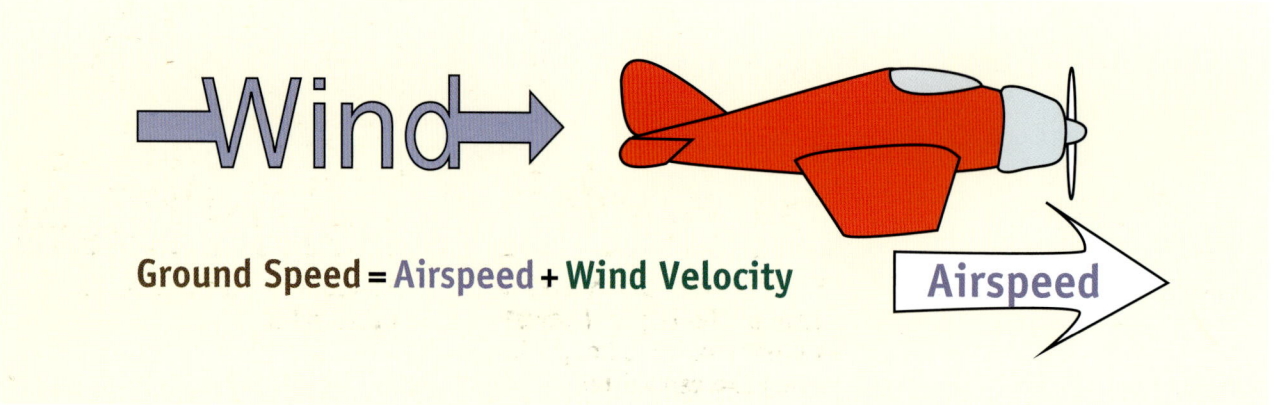

Ground Speed = Airspeed + Wind Velocity

Diamondhead Canard

A good canard with an elegant folding method that starts with a 30° angle. Begin white-side up. Prepare for some difficult origami.

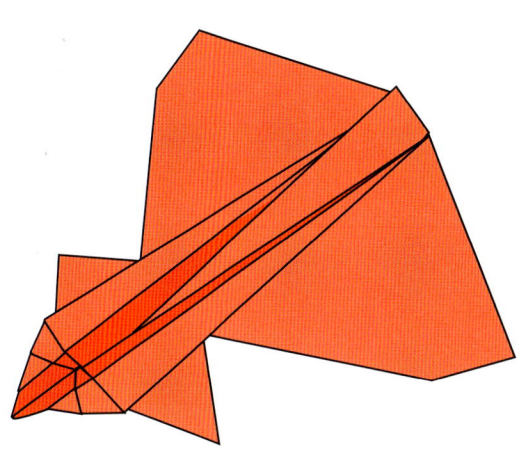

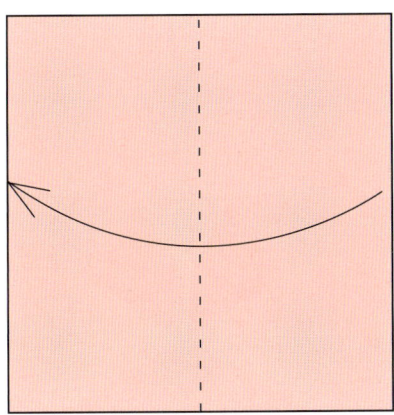

1. Valley fold in half.

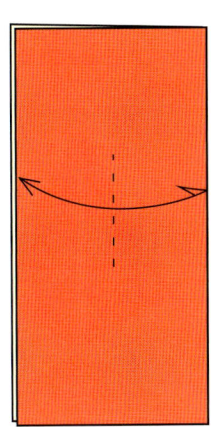

2. Crease lightly partway down the middle.

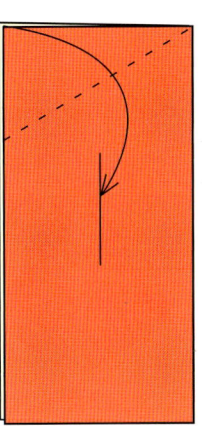

3. Valley fold so that the corner lies on the crease you just made.

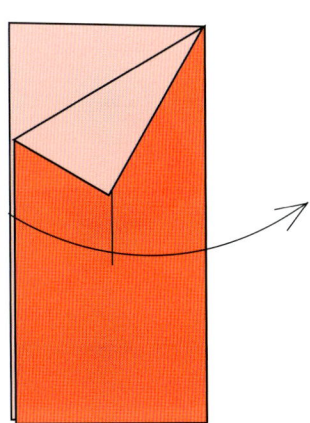

4. Unfold along the center.

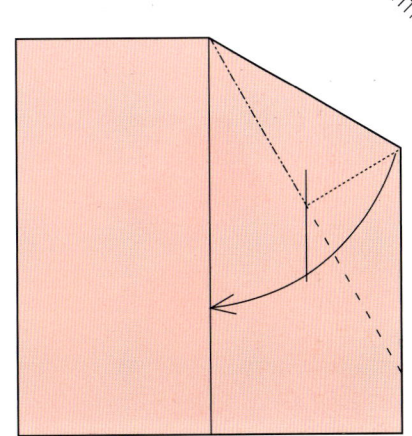

5. Valley fold along the raw edge of the flap underneath, so that the folded edge lies along the centerline.

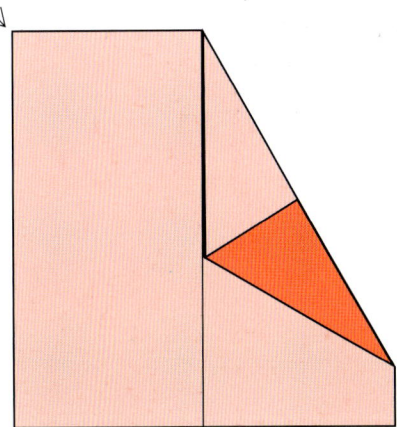

6. Repeat steps 2–5 on the other side.

Canards

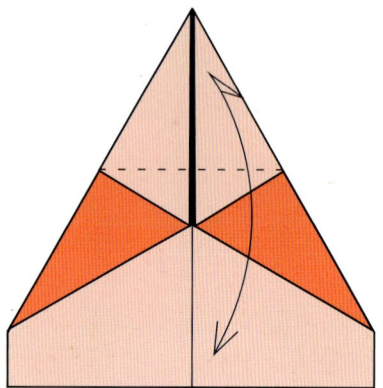

7. Valley fold the top lighter diamond at its widest point and unfold.

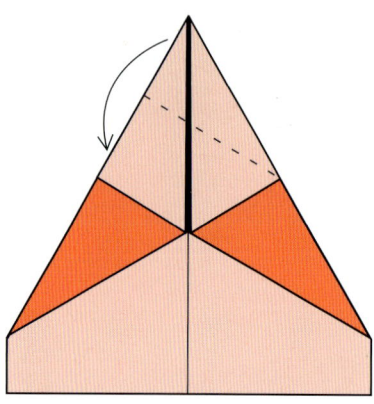

8. Valley fold the top from the corner of the lightly colored diamond perpendicular to the opposite folded edge. The flap should double back on itself when done.

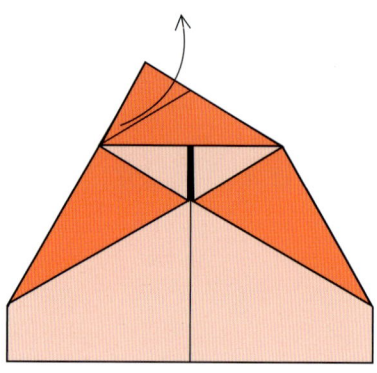

9. Unfold.

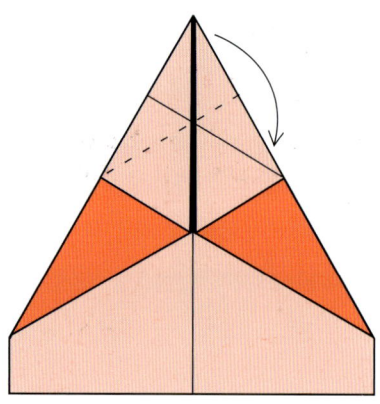

10. Repeat on the other side.

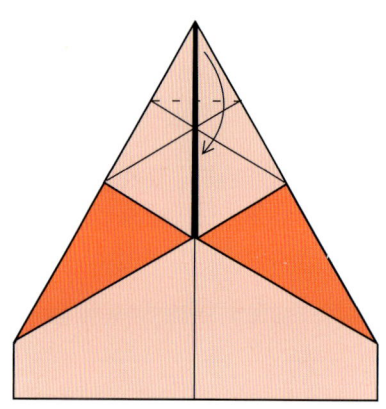

11. Valley fold at the top of the two creases just made, perpendicular to the center.

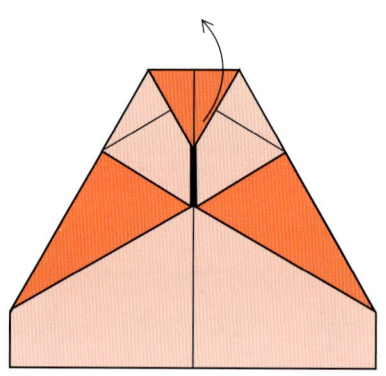

12. Unfold.

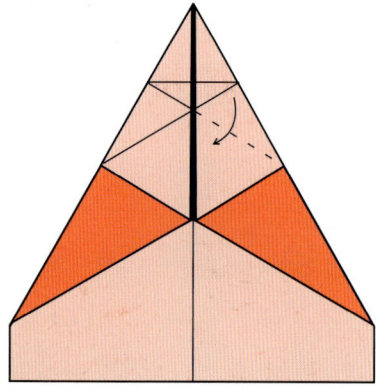

13. Here's where the fun begins. Valley fold along the crease. The top will not lie flat.

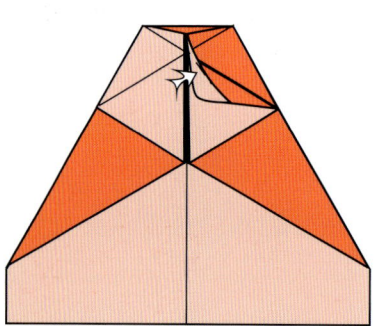

14. Push on the inside of the flap and create a mountain fold so that the paper inside lines up on the centerline.

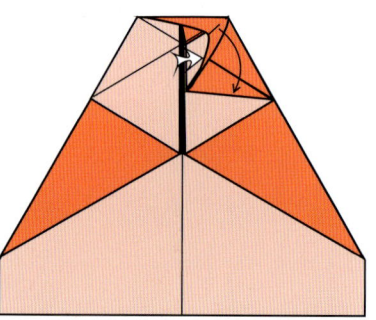

15. Continue to push paper outward as you continue to flatten the front.

70

Diamondhead Canard

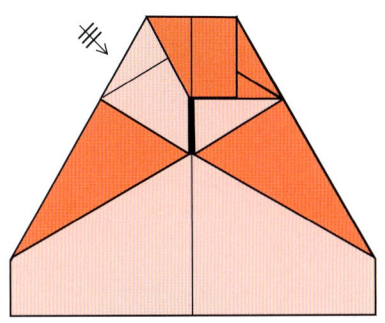
16. Repeat steps 12–15 on the other side.

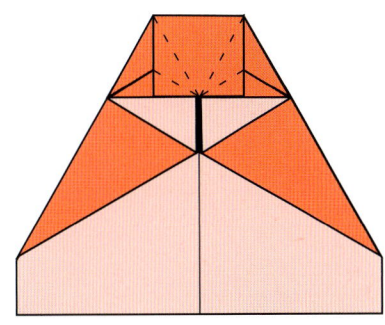
17. Crease with valley folds as shown.

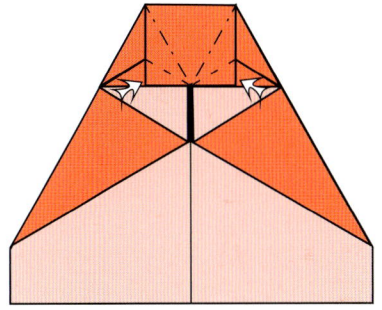
18. Reverse fold in and out.

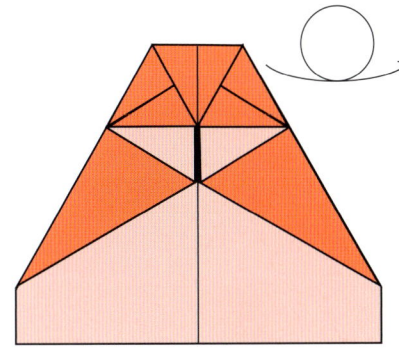
19. Turn over.

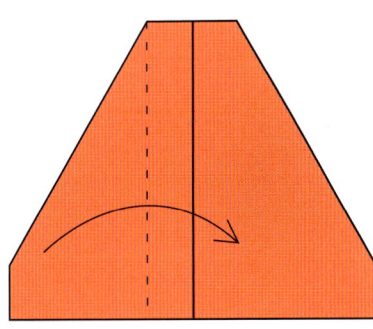
20. Valley fold.

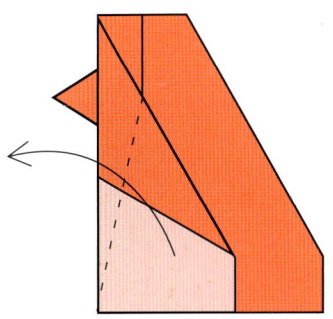
21. Valley fold back out. Note that the top of the fold hits the centerline.

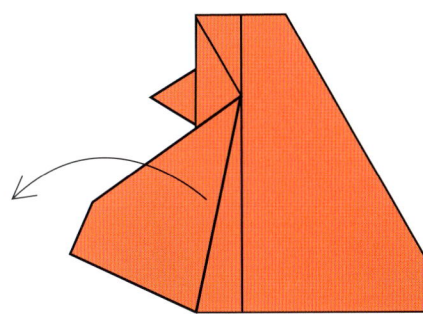
22. Unfold to step 12.

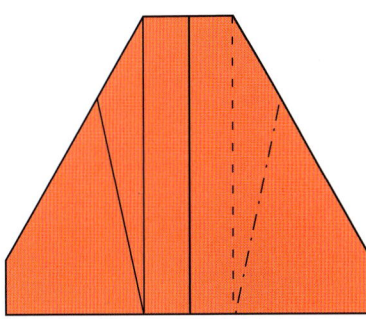
23. Repeat steps 20–22 on the other side.

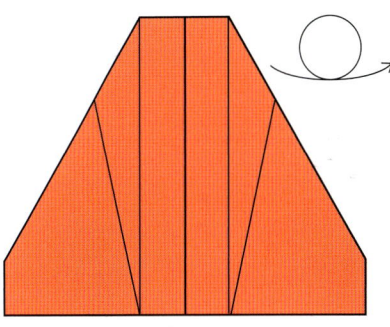
24. Turn over.

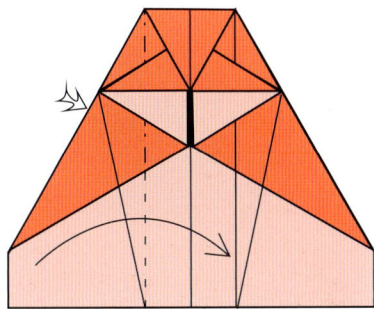
25. Reverse fold the side using the first of the creases made.

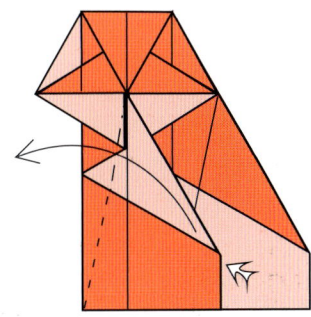
26. Reverse fold back out along a preexisting crease.

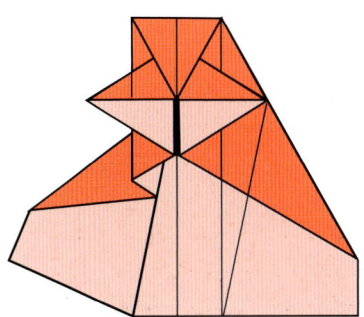
27. Repeat on the other side.

Canards

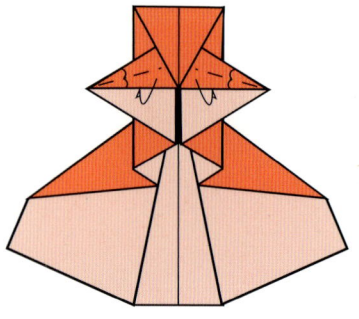

28. Mountain fold the colored paper in the canard in half.

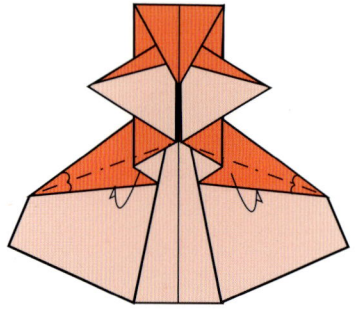

29. Narrow the main wing similarly.

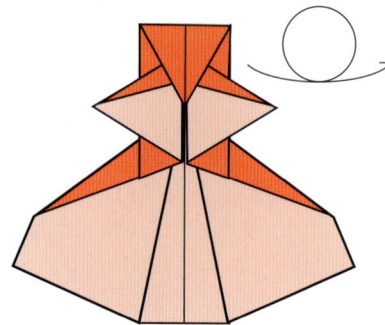

30. Turn over.

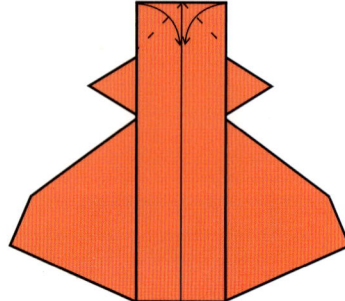

31. Valley fold the front to make a nice, pointy nose.

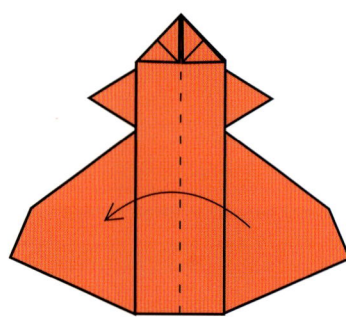

32. Valley fold in half.

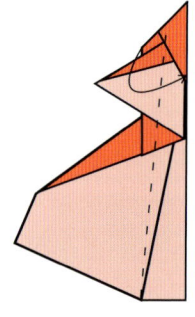

33. Valley fold the wings so the point shown meets the bottom. Repeat on the other side.

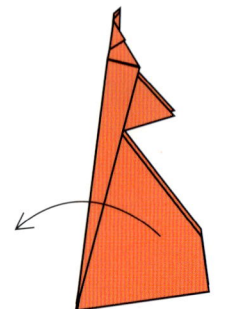

34. Unfold the wings.

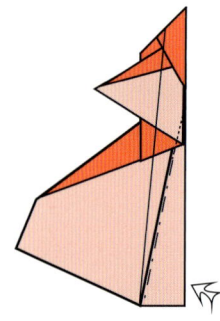

35. Reverse fold the tail along the folded edge.

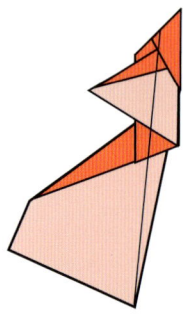

That's it! Open up and you're done.

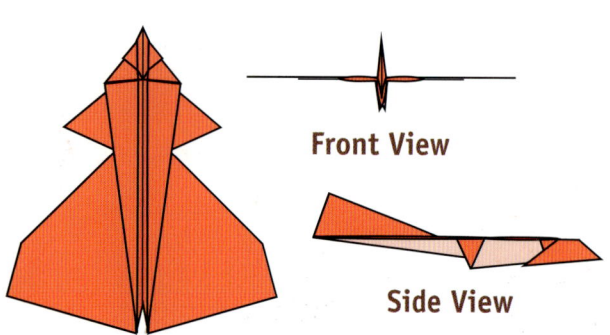

Top View

Front View

Side View

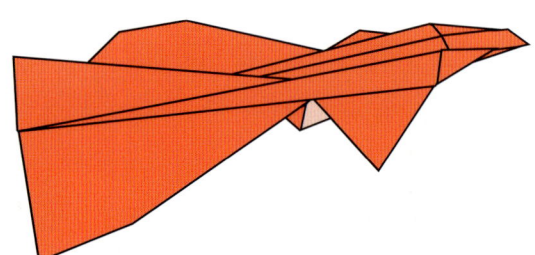

With a bit of elevator, Diamondhead Canard will fly straight. Other than that, it will make a good dart. I hope you've enjoyed folding it, as it's one of my favorites.

Twin Tail

I always wanted to make a paper airplane that looked like a real one—with straight wings in front with a true empennage. I never really succeeded with anything that flew, but came close with this.

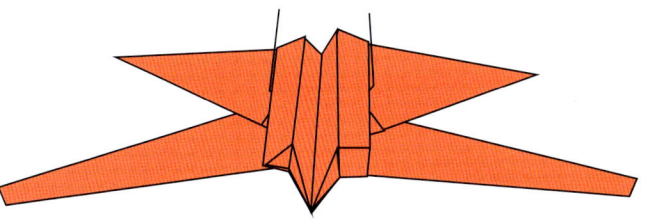

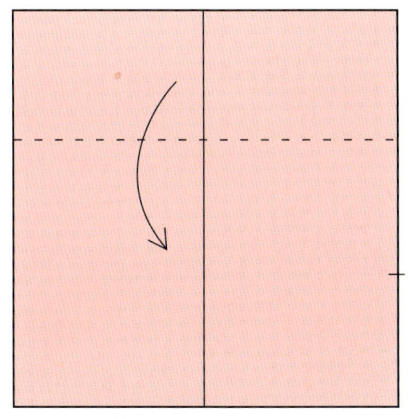

1. Valley fold one third of the way down.

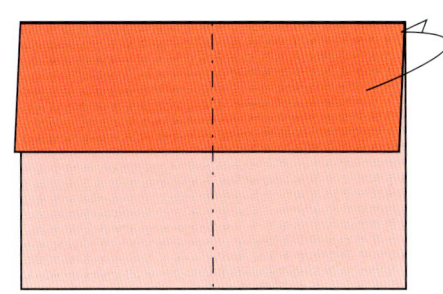

2. Mountain fold in half.

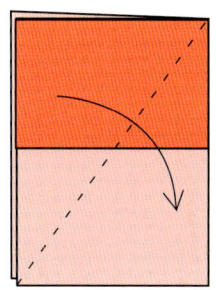

3. Valley fold from the bottom corner to the top middle.

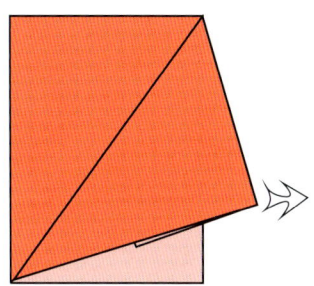

4. Pull out the loose paper.

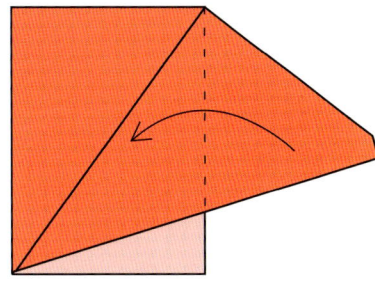

5. Valley fold the flap back to the left.

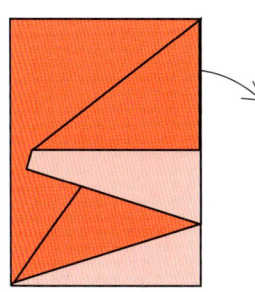

6. Unfold the centerline.

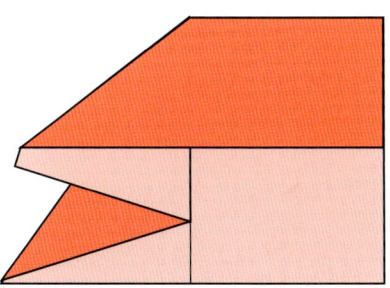

7. Repeat steps 3–5 on the other side.

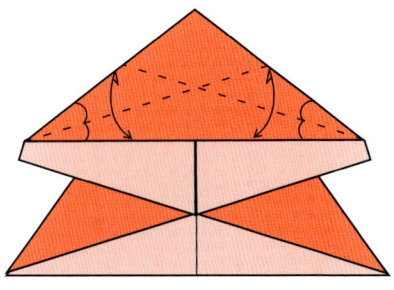

8. Valley fold the angle bisectors and unfold.

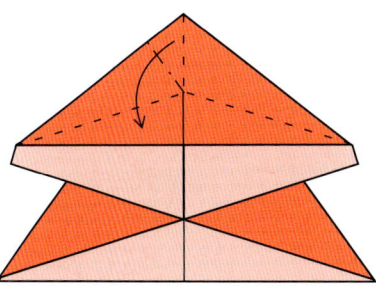

9. Rabbit ear the top.

Canards

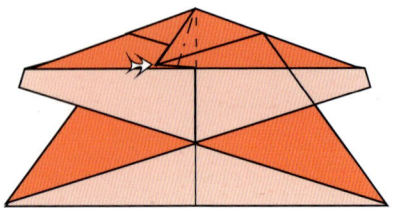

10. Squash fold.

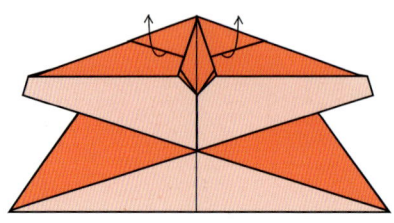

11. Pull out some loose paper.

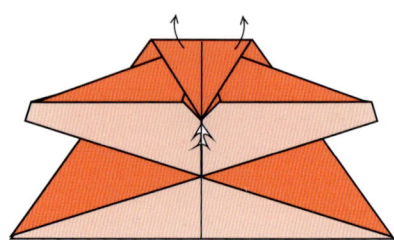

12. Pick the top flap up while holding down underneath, and perform a spread sink fold.

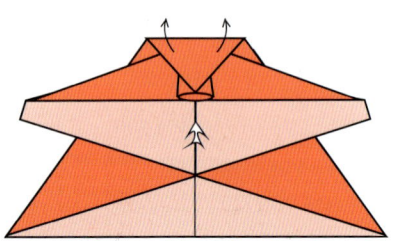

13. The spread sink in progress. Flatten the paper.

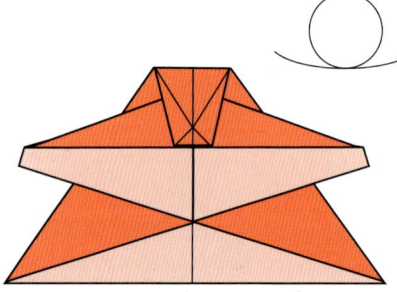

14. Turn over.

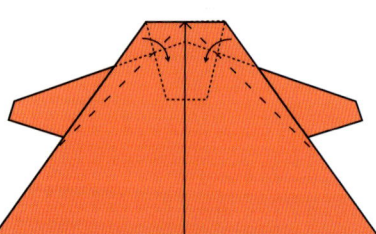

15. Valley fold the front corners as far as they will go.

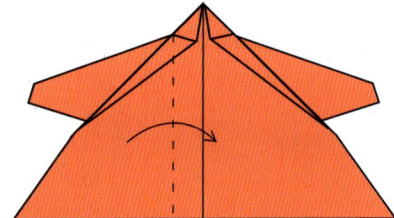

16. Valley fold the left flap as far as it will go.

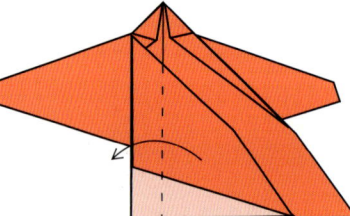

17. Valley fold back at the centerline.

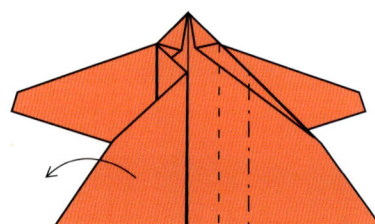

18. Unfold, and repeat on the other side.

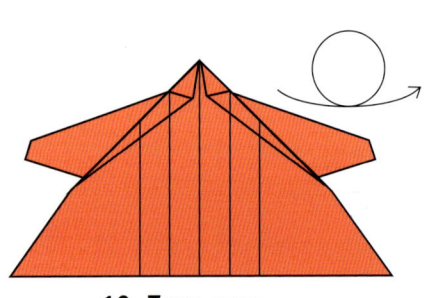

19. Turn over.

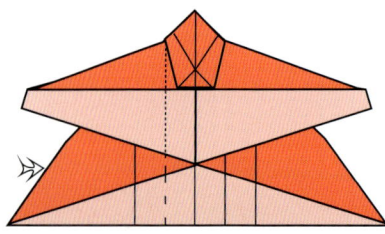

20. Inside reverse fold along an existing crease.

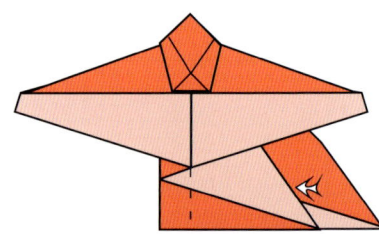

21. Reverse fold back over.

Twin Tail

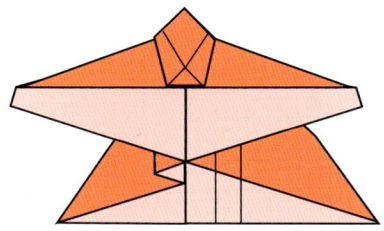

22. Repeat steps 20 and 21 on the other side.

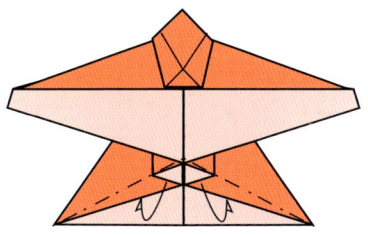

23. Mountain fold some paper on the back wings behind.

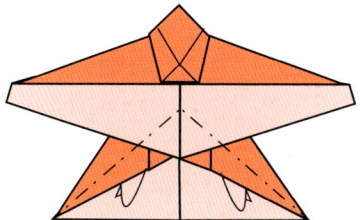

24. Mountain fold some more. This will shift the weight up to the front.

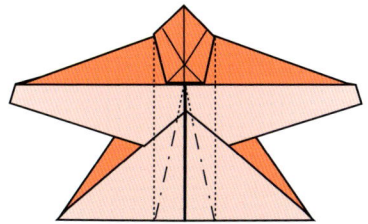

25. Reverse fold out the two tails. Get as much tail as you can.

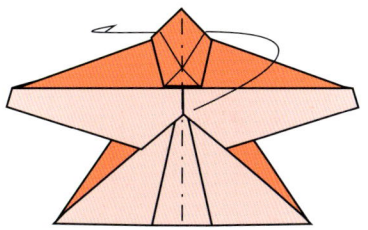

26. Mountain fold in half.

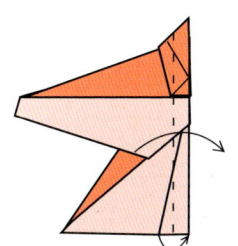

27. Finish up by valley folding the wings.

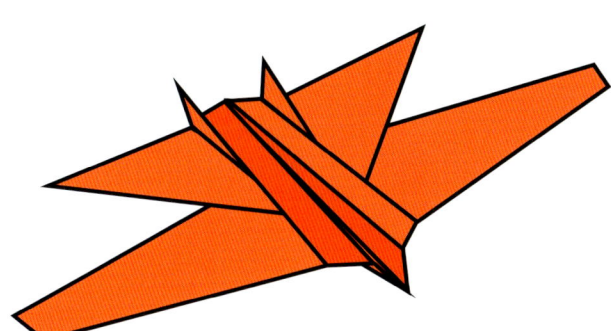

The Twin Tail, which looks like an Art Deco aircraft, is made similarly to the previous airplane. Re flying, hmm. With its high-aspect-ratio wings it makes a poor dart. On the other hand, it has less wing area than most of the gliders, increasing its wing loading (more weight on less area), so it doesn't glide as well as the gliders. Given a hard throw, it will do a loop or a tight turn, and then land with a semi-pathetic glide. It does look cool while doing it, though.

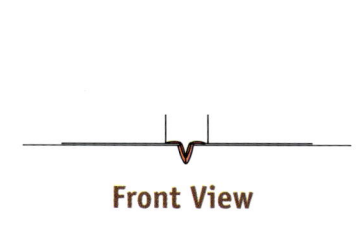

Front View

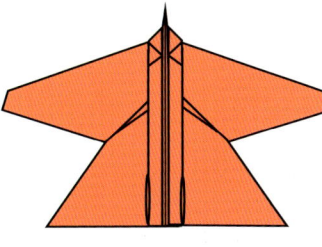

Top View

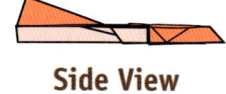

Side View

> *The higher we soar, the smaller we appear to those who cannot fly.* — *Friedrich Wilhelm Nietzsche*

The Drop Zone

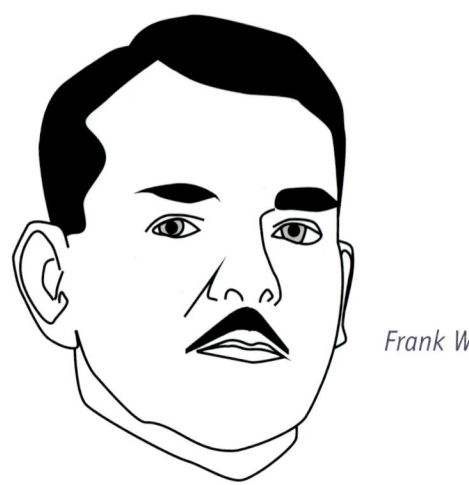

Frank Whittle

Nothing epitomizes modernity like jets. Phrases such as "Jet Age" and "jet set" conjure up images of speed, comfort, and world travel. But jet propulsion might have never come to us if not for the machinations of a brilliant Royal Air Force pilot named Frank Whittle.

Whittle was born in Coventry, England, in 1907, the son of a mechanic and engineer, from whom he picked up engineering skills early in life. But Whittle didn't want to follow in his father's footsteps; he was determined to be a pilot. To that end he applied to the RAF, enlisting as an aircraft mechanic's apprentice. Fortunately for him, his keen intellect and meticulous aircraft-modeling skills came to the attention of his superior officer, who recommended him for officer training.

This was an incredible opportunity for Whittle, as the officer corps was mostly composed of aristocrats. The transition was very difficult, as he was a commoner, but he was able to pursue his dream of being a pilot, and quickly learned to fly the fighters of the day. Indeed he was known for his daring low-level aerobatics, so much so that he was nearly court-martialed.

Every officer candidate was required to write a thesis, and Whittle wrote his on the limitations of propeller-driven aircraft at high speed and altitude. He envisioned a hybrid engine in which a piston engine compressed air that was then ignited and exhausted out the back, basically like an afterburner. He refined his design to use a turbine engine, and presented his ideas to the British Air Ministry in 1929, but an analysis by a colleague dissuaded the military, which failed to evince any interest in Whittle's idea.

He was able to acquire partners and funding from venture capitalists to pursue his engine design, and in 1935 set up Power Jets Ltd. to begin development. Funding was undependable, however, and the design presented numerous developmental difficulties. By 1939 they had a workable prototype, but had nearly run out of funds. Whittle's health suffered tremendously during this time, and he suffered a nervous breakdown.

The period of difficulty came to an end when Air Ministry officials watched the engine run for twenty minutes with no troubles. The ministry immediately injected funds into the company, and ordered an aircraft built to serve as a test bed for the new engine. On April 12, 1941, the first British jet aircraft lifted into the skies over Gloucester. With its experimental engine it quickly outperformed the Supermarine Spitfire, which was powered by the most advanced piston engine of its day. The Jet Age was born.

The first jet engine

Drop Zone

The vertical stabilizers that drop down from the wings make this a distinctive airplane that will do excellent stunts and give nice glides. Begin white-side up.

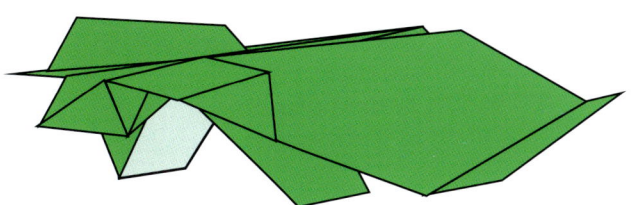

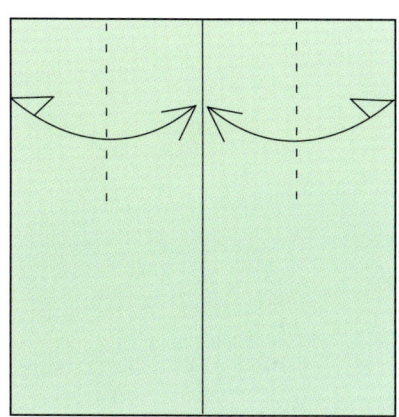

1. Valley fold the raw edges at the sides into the center. Crease only in the top half, and unfold.

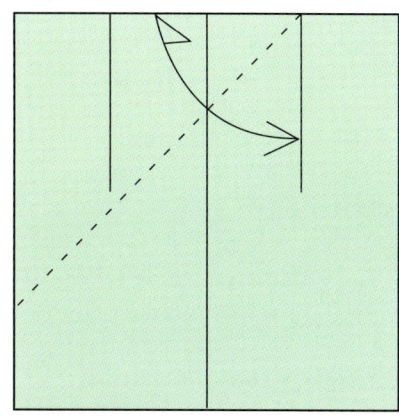

2. Fold the top raw edge to the crease you just made. Crease well, unfold, and repeat on the other side.

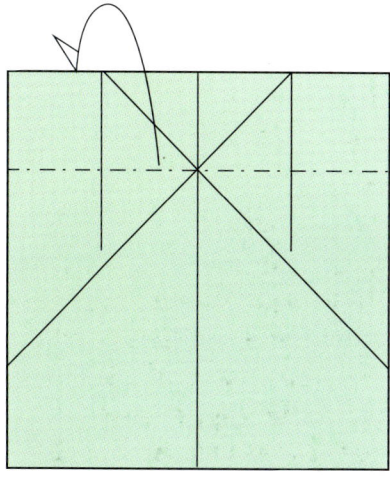

3. Mountain fold where the two diagonals meet. Crease well, and unfold once again.

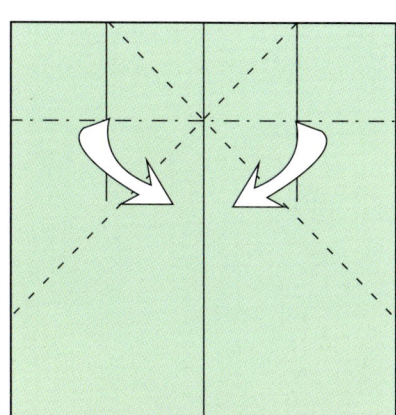

4. Collapse the top by folding along the creases you made in the preceding steps.

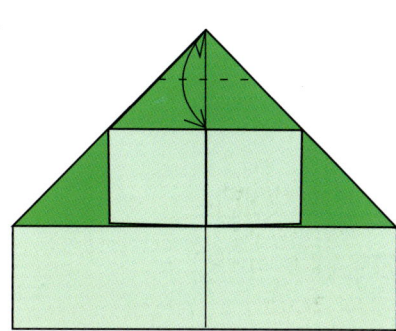

5. Fold the top point down to meet the raw edge beneath it. Crease well and unfold.

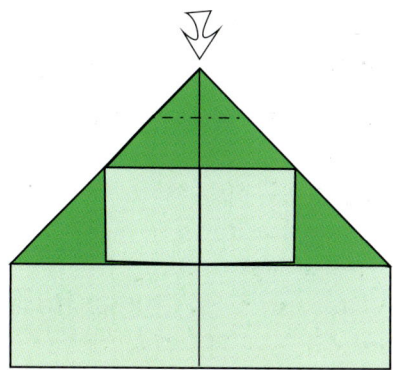

6. Sink the top.

The Drop Zone

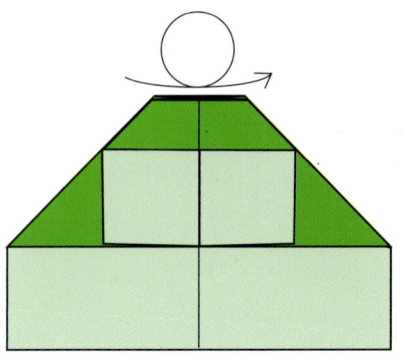

7. Turn over.

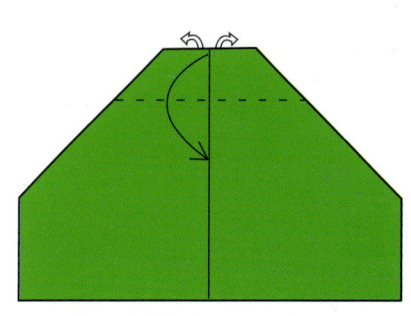

8. Valley fold the top down while opening up the flaps in the middle.

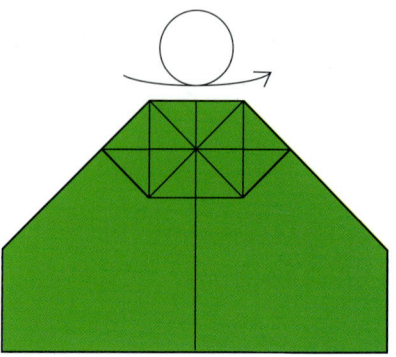

9. Turn over.

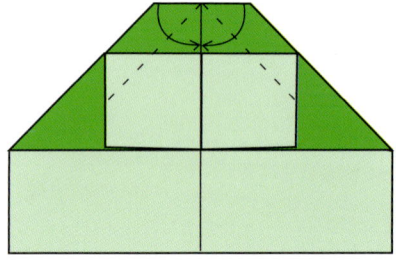

10. Narrow the upper layer using diagonal valley folds.

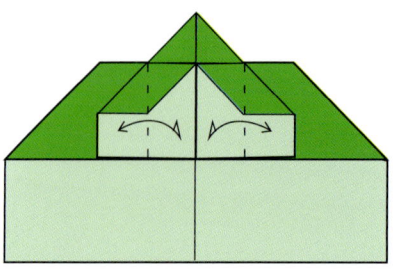

11. Valley fold the ventral stabilizers and unfold.

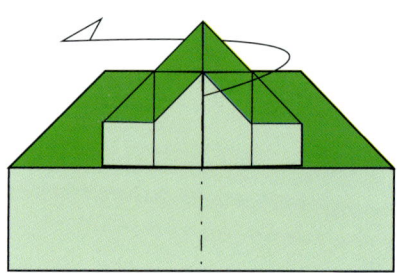

12. Mountain fold in half.

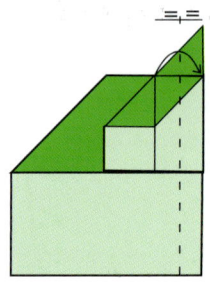

13. Valley fold the wings by dividing the nose of the airplane in half. This gives nice, big wings. But remember: it's not the size of your wings that counts, it's how you use them.

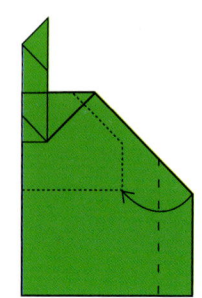

15. To finish the Drop Zone, fold a pair of dorsal vertical stabilizers. The tips of these should just reach the ventral stabilizers below, though the actual placement is not critical.

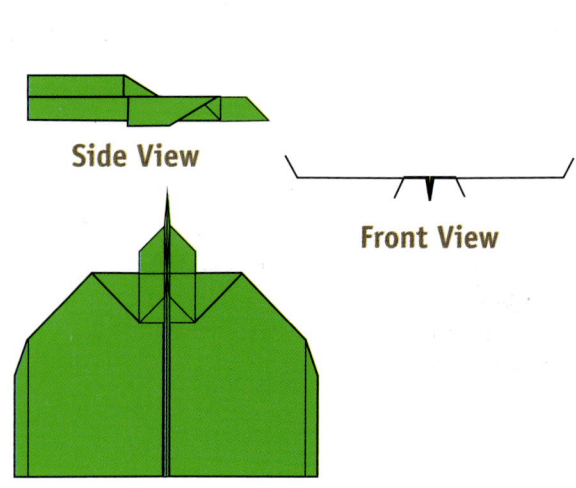

Side View

Front View

Top View

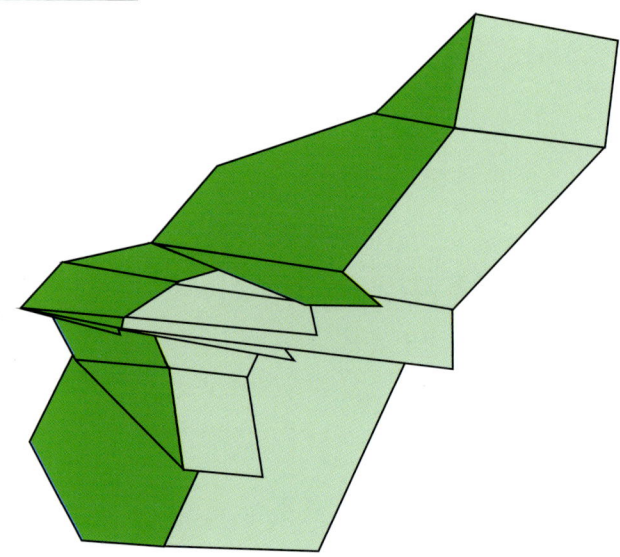

Bird-base Glider

As the name suggests this airplane starts with a Bird Base—well, sort of. It also has drop-down stabilizers, and was one of the first aircraft in this collection. Begin with step 5 of the Bird-base Fighter.

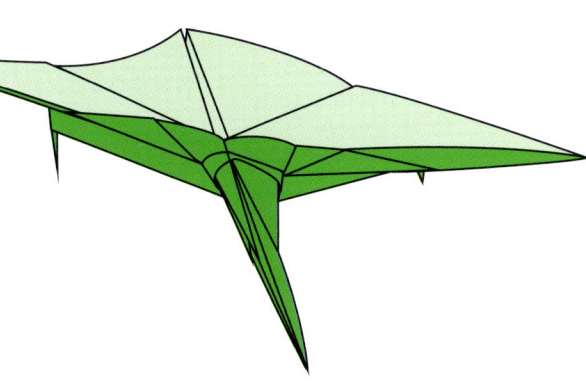

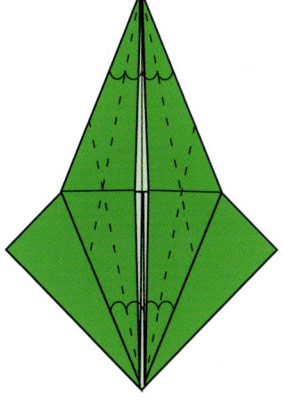

1. Crease the angle bisectors top and bottom.

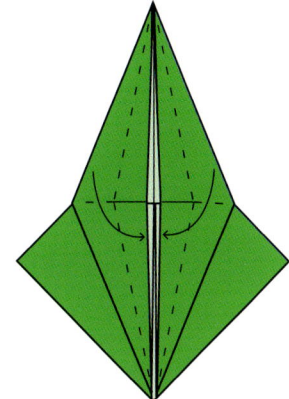

2. Rabbit ear to the center.

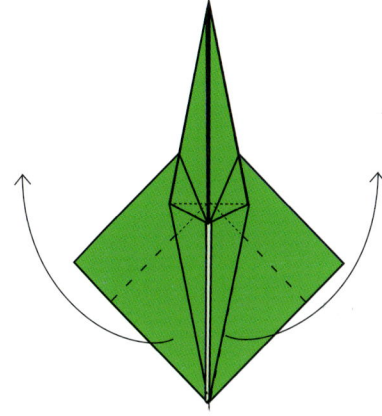

3. Swivel the flaps upward.

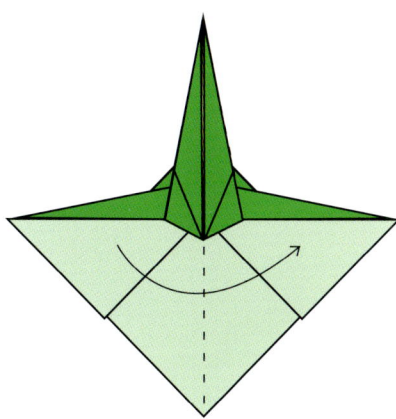

4. Valley fold the model in half.

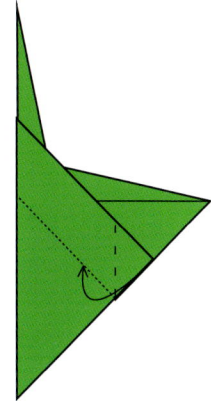

5. Valley fold a stabilizer by lining up the rear raw edge with the flap hidden underneath. You can visualize this by holding the model up to a light.

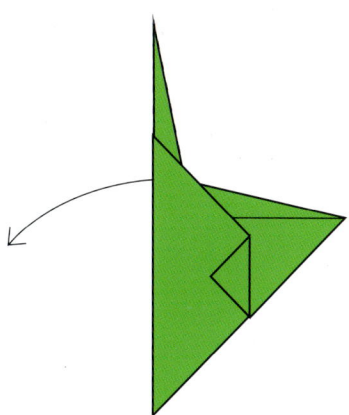

6. Unfold.

The Drop Zone

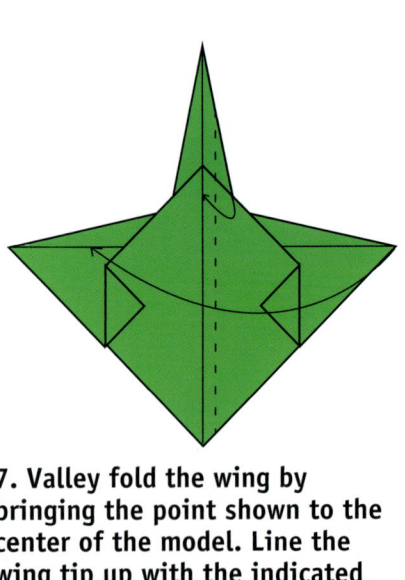

7. Valley fold the wing by bringing the point shown to the center of the model. Line the wing tip up with the indicated crease and fold. Be careful, as there's a lot of thickness there.

8. Mountain fold along the centerline.

9. Mountain fold the second wing to match the first.

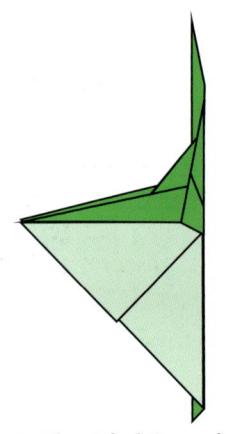

10. The Bird Base is ready to fly.

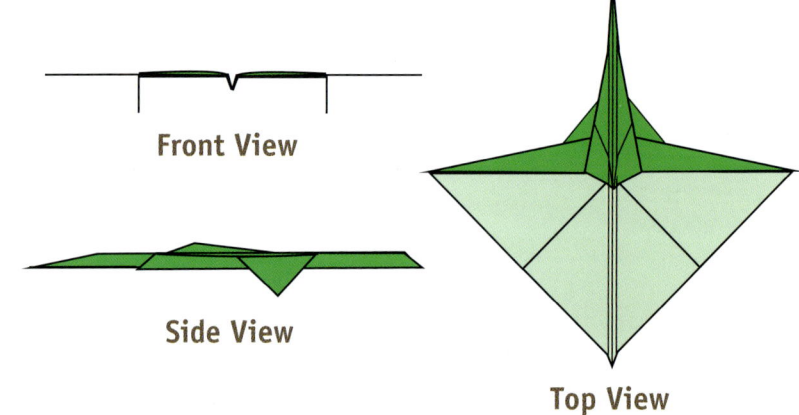

Front View

Side View

Top View

BBG will glide nicely when given a gentle throw.

The strength of the turbulence is directly proportional to the temperature of your coffee. — Gunter's Second Law of Air Travel

80

Tail Dragger

This is a good example of weight-forward airplane design. It started as an attempt to use a blintzed fish base as the front of an airplane, but was too heavy in the rear. Concentrate a little paper over in front and it works like a charm. Besides, the folded front winds up making an interesting-looking nose. Begin white-side up.

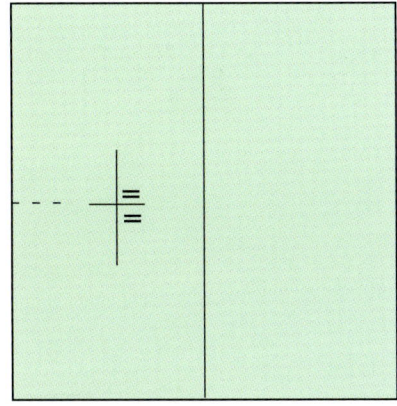

1. Lightly crease the raw edge halfway down.

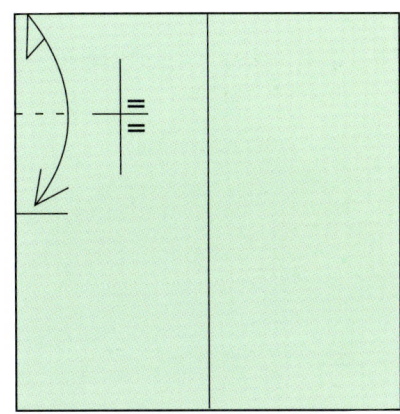

2. Make another crease one quarter of the way down.

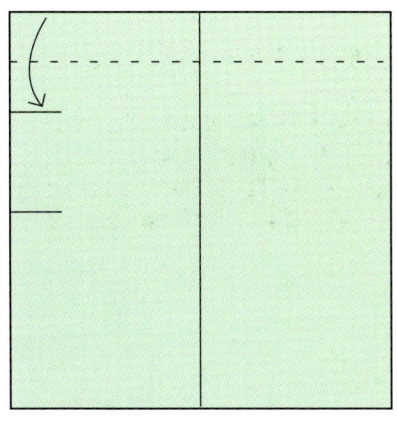

3. Valley fold one eighth down. This is another example of folding the front to put the CoG forward.

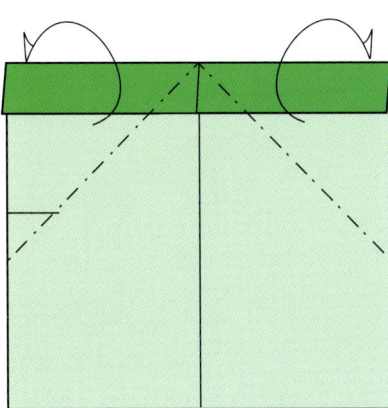

4. Mountain fold the corners so that the top folded edge lies on the centerline.

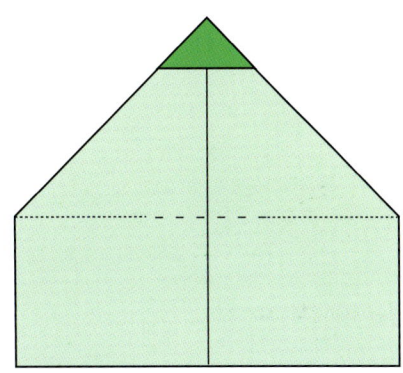

5. Make a light crease across the bottom of the upper triangle.

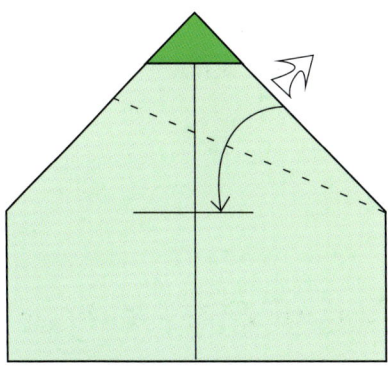

6. Valley fold so that the top folded edge lies on the crease you made in step 5. Allow the layer behind to flip out.

The Drop Zone

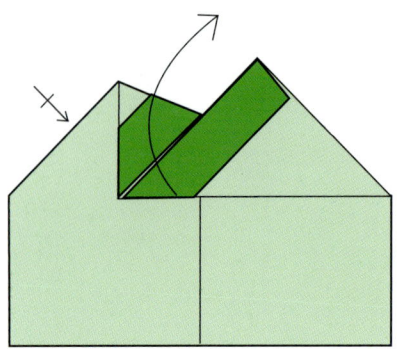

7. Unfold, and repeat on the other side.

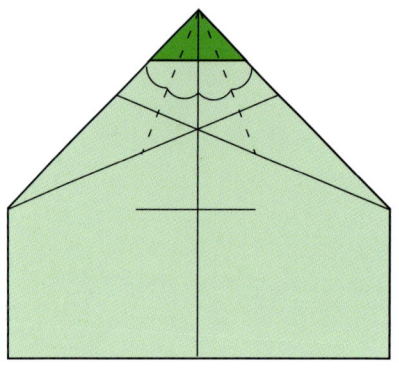

8. Crease along the angle bisectors.

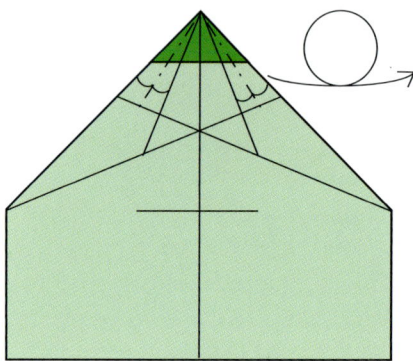

9. Crease again along the angle bisectors. Turn over.

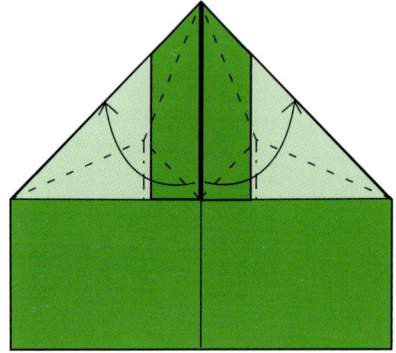

10. Rabbit ear both flaps.

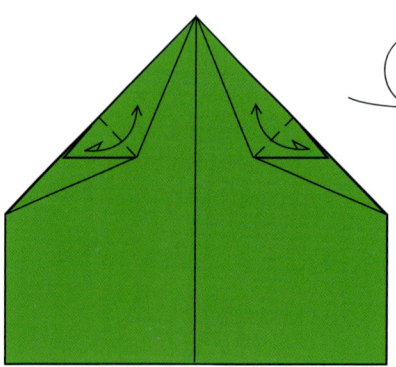

11. Fold the rabbit ears back and forth. Turn over.

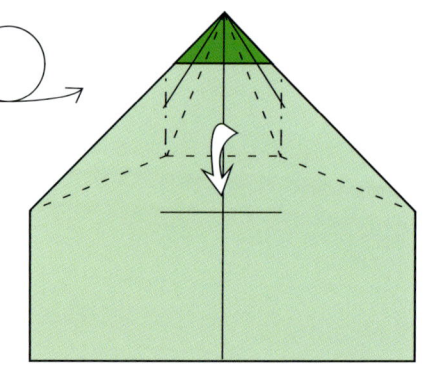

12. Collapse the top along the creases made in the previous steps.

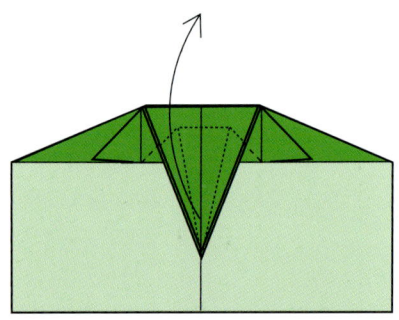

13. Squash fold the top flap upward along the creases you made previously. You get to make some new ones, too.

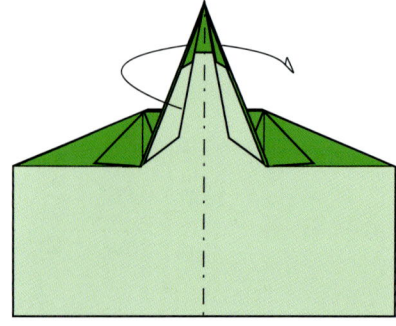

14. Mountain fold the airplane in half.

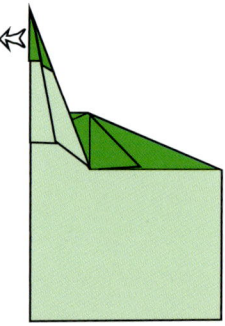

15. Pull out some loose paper from the front.

Any landing you can walk away from is a good one! — Gerald R. Massie

Tail Dragger

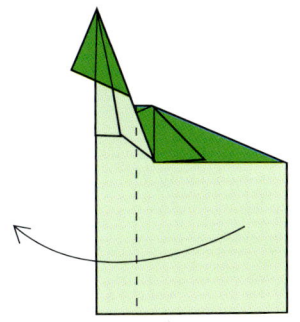

16. Valley fold the wings down.

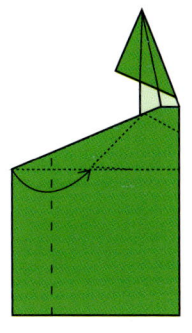

17. Valley fold the upper vertical stabilizers to about where the lower ones sit.

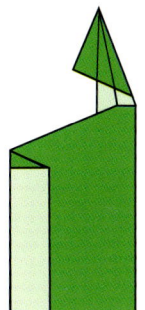

Done.

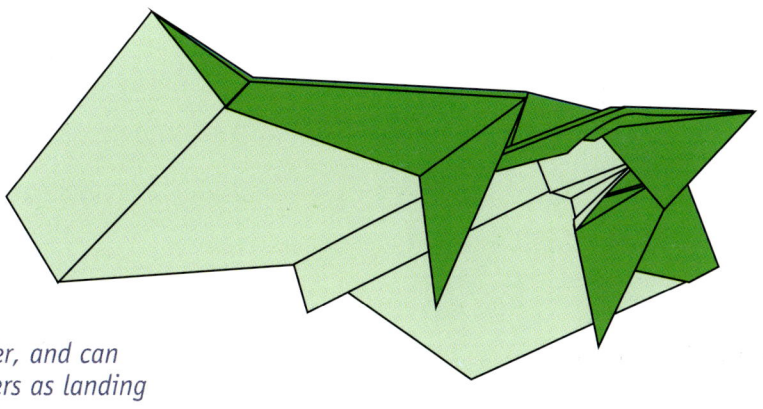

Tail Dragger is also a good glider, and can use its lowered vertical stabilizers as landing gear, thus giving it its namesake.

Top View

Side View

Front View

It is not only fine feathers that make fine birds. — Aesop

83

Cessna 172

Tail Dragger is so named because the vertical stabilizers also act as landing gear, and its tail sits on the ground at rest. Antique aircraft, such as the Boeing Stearman, all land with the tail on the ground and are therefore affectionately known by pilots as tail draggers, though our good pals at the FAA say the landing gear is in the conventional configuration. Landing these aircraft can be quite tricky, and it's said that they're flown from the moment they start rolling until they reach a full stop, whether they're on the ground or not.

Most modern airplanes have tricycle gear, with one wheel up front and two in back, like the Cessna 172 pictured here. The engineering of the front wheel was somewhat difficult, so aircraft with nose wheels didn't come into wide use until after World War II. However, such airplanes are considerably easier to land than tail draggers, and are now in wide use. In its advertising, Cessna labeled its first tricycle-gear aircraft *landomatics*, and the moniker stuck.

Good pilots love a challenge, and tail draggers present one, giving them some enduring popularity. Despite the fact that special training is needed to land an aircraft with conventional gear, they're still quite popular, and there are many vintage aircraft still flying and meticulously maintained by their owners. Many people who elect to build their own aircraft construct tail draggers, some because of the vintage appearance, and some because they like the challenge of landing a more demanding airplane.

Not pictured here are wheel covers, or spats, which many pilots put on the landing gear to make it more streamlined. I have them on my aircraft because I think they look better. Some take them off, as they can get damaged when landing on a rough surface. Many aircraft are outfitted with floats to land on water, although this is quite difficult and requires a special rating. Others are outfitted with skis, to land on winter snow in the northern states.

Boeing Stearman

Gremlin

The first airplane design I came up with using the Preliminary Base. Begin with step 4 of the Bird-base Fighter.

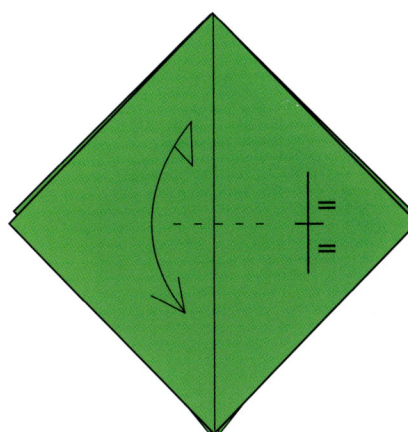

1. Fold the bottom point up to the top, crease lightly along the center, and unfold.

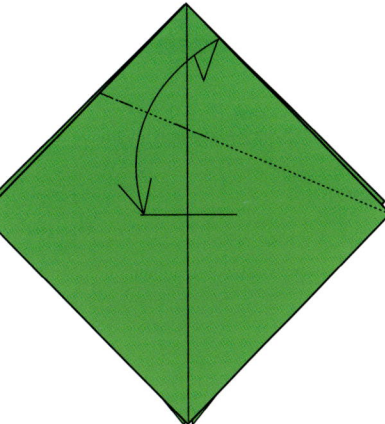

2. Fold so that the multiple folded edges at the top touch the center crease. Crease very lightly at the midline and the top edge, and unfold.

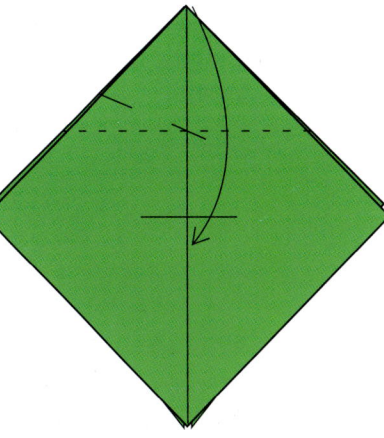

3. Valley fold where the crease made in the previous step meets the middle.

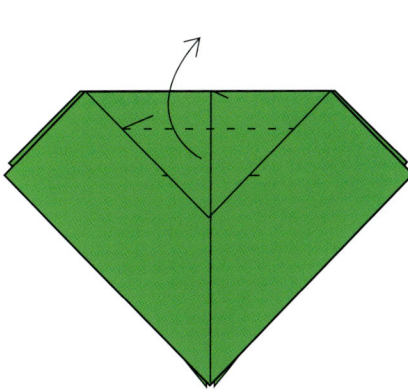

4. Valley fold the flap back up where the crease made in step 2 meets the edge.

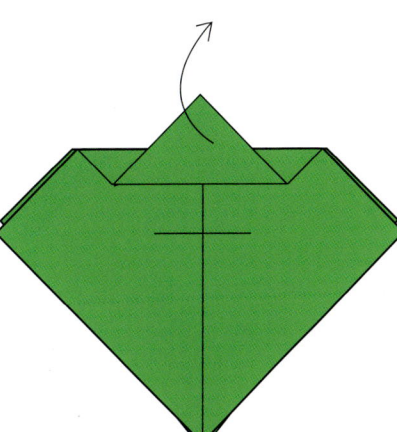

5. Unfold.

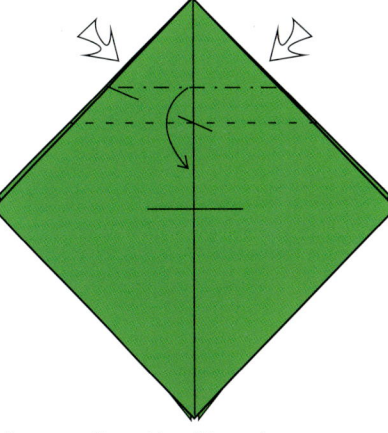

6. Now this gets fun. You'll make a spread-squash, which is sort of like a slanted lovers' knot. Valley fold on the lower crease while mountain folding on the upper one—but on the top layer only.

85

The Drop Zone

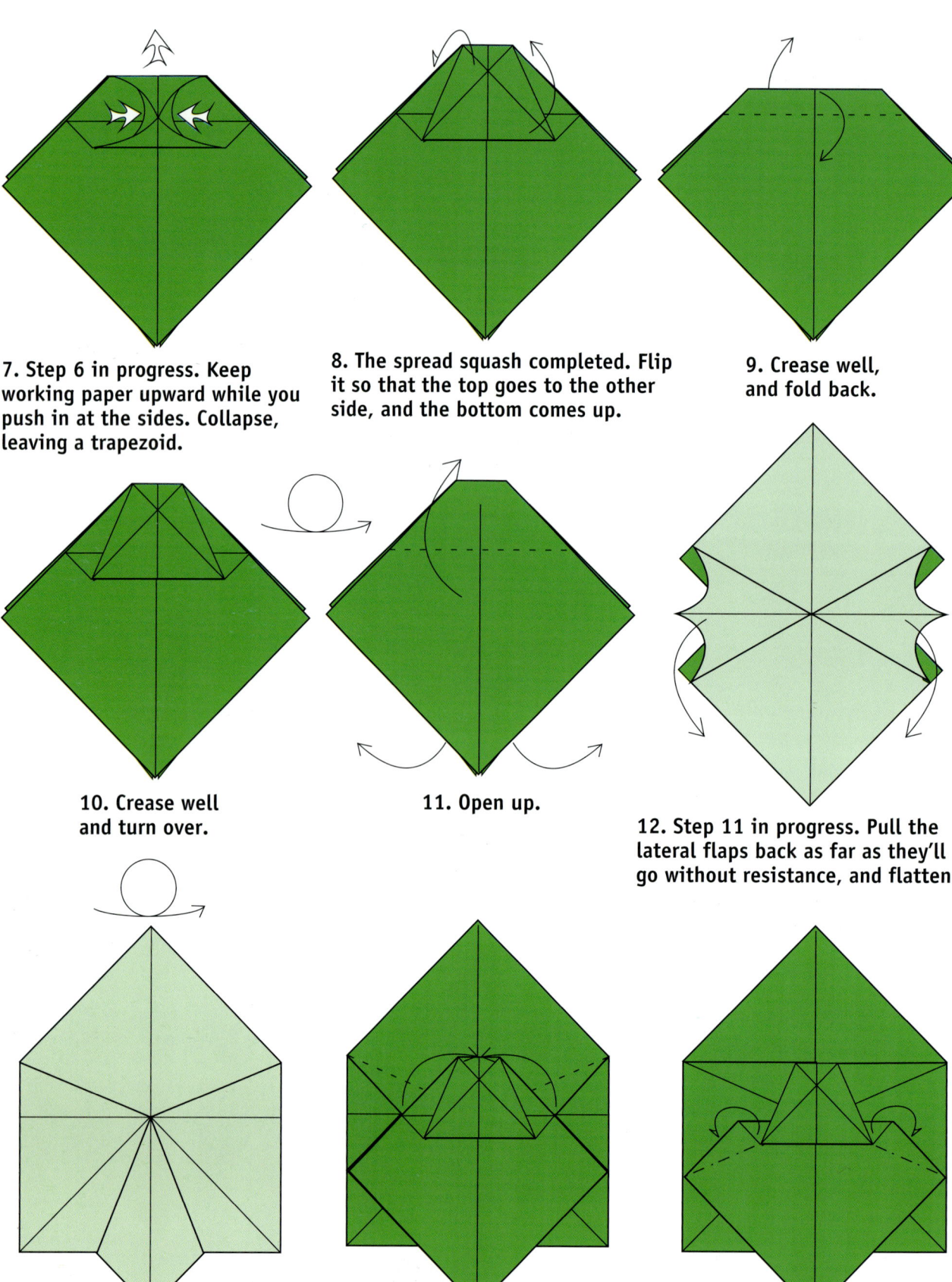

7. Step 6 in progress. Keep working paper upward while you push in at the sides. Collapse, leaving a trapezoid.

8. The spread squash completed. Flip it so that the top goes to the other side, and the bottom comes up.

9. Crease well, and fold back.

10. Crease well and turn over.

11. Open up.

12. Step 11 in progress. Pull the lateral flaps back as far as they'll go without resistance, and flatten.

13. Turn over.

14. Swivel two flaps into the center.

15. Mountain fold the indicated flaps behind.

86

Gremlin

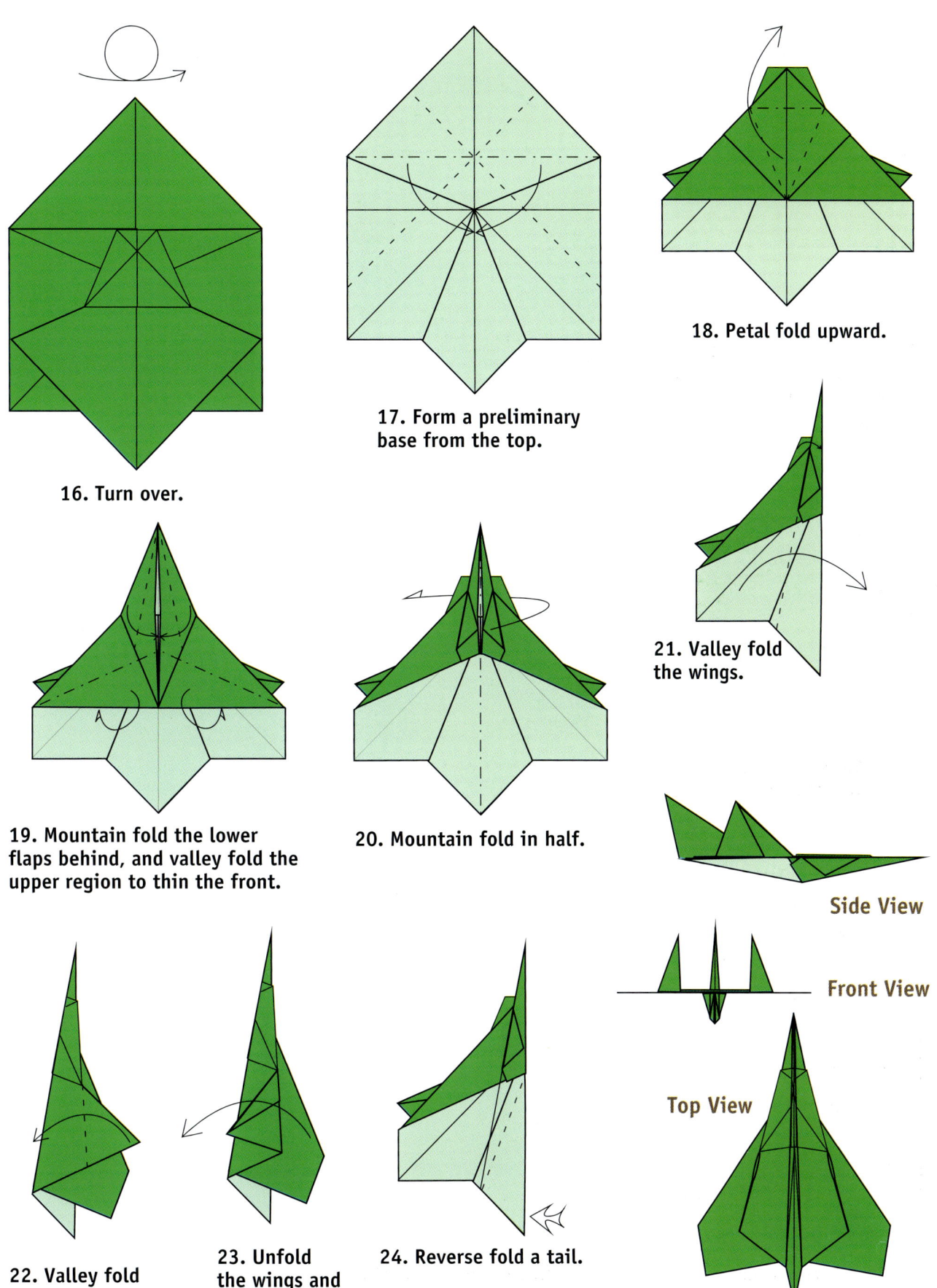

16. Turn over.

17. Form a preliminary base from the top.

18. Petal fold upward.

19. Mountain fold the lower flaps behind, and valley fold the upper region to thin the front.

20. Mountain fold in half.

21. Valley fold the wings.

22. Valley fold vertical stabilizers.

23. Unfold the wings and stabilizers.

24. Reverse fold a tail.

Side View

Front View

Top View

87

Brancusi Animals

Although Constantin Brancusi had little to do with aviation, he did invent abstract sculpture.

Brancusi was born in a small village in Romania's Carpathian Mountains, an area with a rich tradition in woodcarving. Though he was the son of poor peasants, young Brancusi was sufficiently talented to earn admission into the Craiova School of Crafts and, later, the Bucharest School of Fine Arts, where he excelled at sculpture.

Brancusi then made the 1,200-mile journey to Paris (by foot, legend has it) where he worked briefly in the studio of the famous sculptor Rodin. As he left he said, "Nothing can grow under big trees."

Branching out on his own, he began sculpting in a fashion very different from that of his contemporaries. At the time, most artists molded their works out of clay or plaster, then had them cast in metal. Instead, Brancusi began carving, and rather than make literal representations of his subjects, made his works simpler and more abstract, trying to show the essence of what he was depicting.

Brancusi's art elicited a very famous lawsuit in 1926. Escorted by Marcel Duchamp, a number of his works had traveled to New York for exhibit.

Constantin Brancusi

But customs officials decided that these abstract works couldn't possibly be art (they classified Brancusi's soaring *Bird in Space* as a kitchen utensil). While art could be imported duty-free, it was decided that a tariff had to be paid on Brancusi's works. Brancusi appealed the decision, and the court case centered on the question: *what is art*? The witnesses included several prominent artists and art critics, and eventually the judge ruled in Brancusi's favor.

We origami artists are especially indebted to Brancusi. Though there are now extremely complex technically advanced designs that reproduce natural subjects with extreme fidelity, all of our works are to at least some degree abstracted. This is even truer in making flying origami; beaks, feet, and other details are not aerodynamic. Thus if I can't imitate the great artist's genius, I can at least steal his ideas.

Never fly in the same cockpit with someone braver than you. — Richard Herman Jr.

Enormously Abstract Heron

A very abstract heron that flies. Begin with step 5 of the Bird-base Fighter.

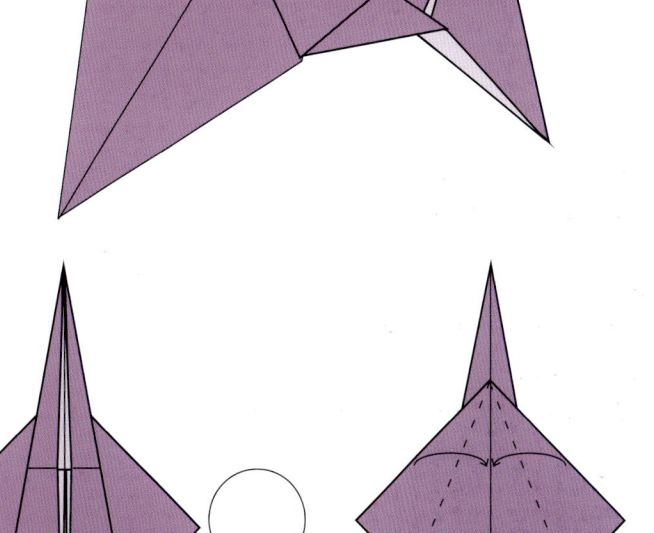

1. Mountain fold to thin down the front.

2. Turn the model over.

3. Valley fold along the angle bisectors.

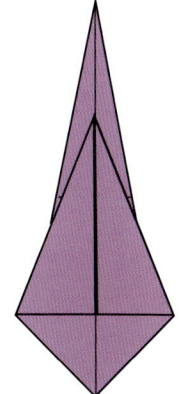

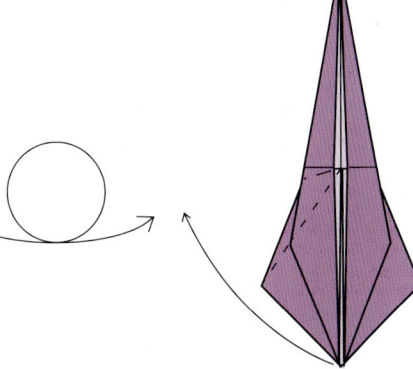

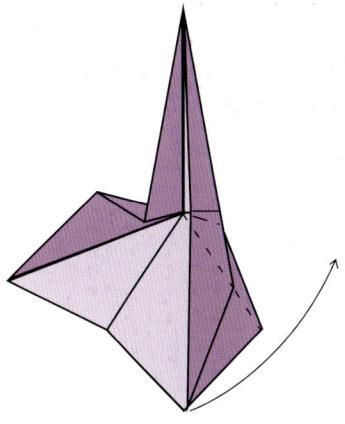

4. Turn over.

5. Swivel out the wing. Note that the valley fold lies along the edge of a layer running underneath.

6. Repeat on the other side.

Brancusi Animals

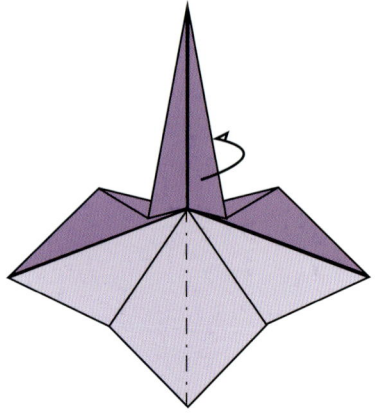

7. Mountain fold in half.

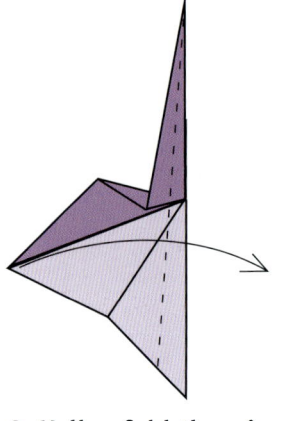

8. Valley fold the wings by lining up the outside and inside edges at the front.

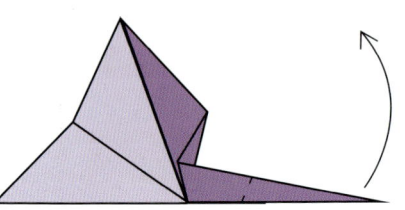

9. You can stop here, and fold the wings by bisecting the angle in front. Or, keep going for a somewhat less abstract version. Reverse fold the head up.

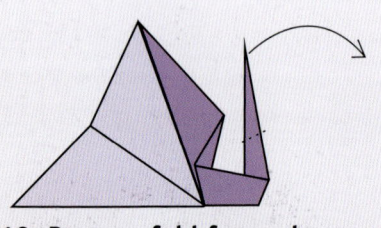

10. Reverse fold forward.

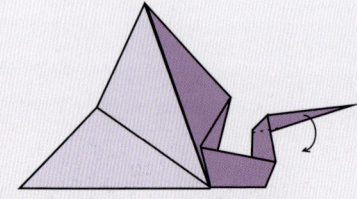

11. Outside reverse fold.

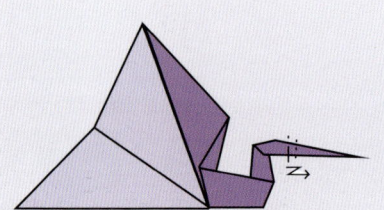

12. Mountain and valley fold to form a beak.

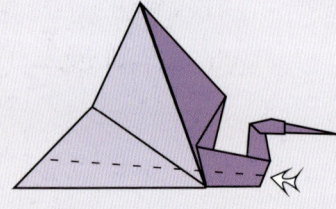

13. When you fold the wings down, squash fold the front.

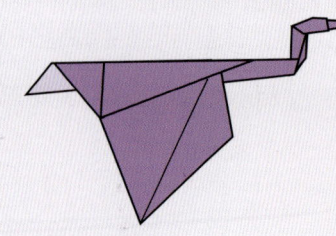

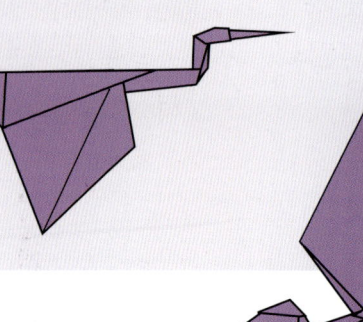

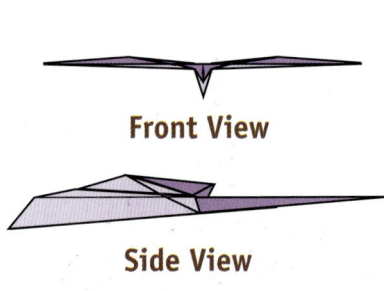

Front View

Side View

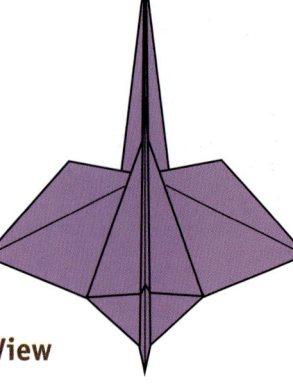

Top View

The more abstract version flies a bit better, but it does look a little more like a bird with its head. Whether you like the abstract or not, it will fly quite well.

Somewhat Less Abstract Canada Goose

A variation of the Heron that also functions quite well. Begin with step 7 of the Heron.

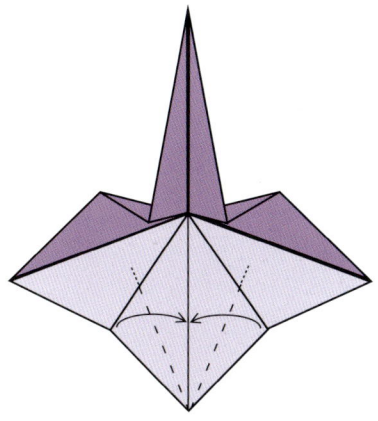

1. Narrow the rear with valley folds. You'll have to move paper from underneath the wings.

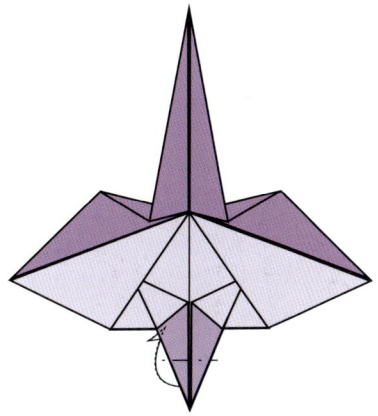

2. Blunt the aft end with a mountain fold.

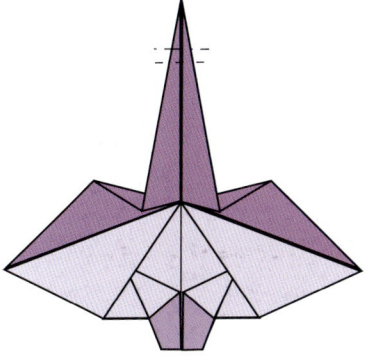

3. Pleat the front.

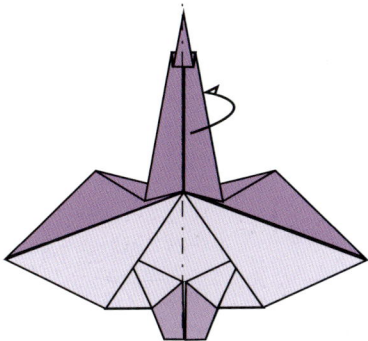

4. Mountain fold the model in half.

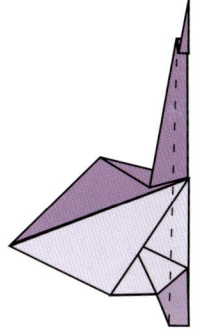

5. Valley fold the wings from the front of the head to the base of the tail.

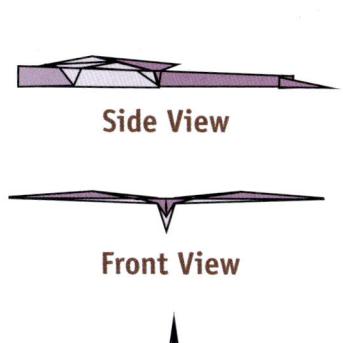

Side View

Front View

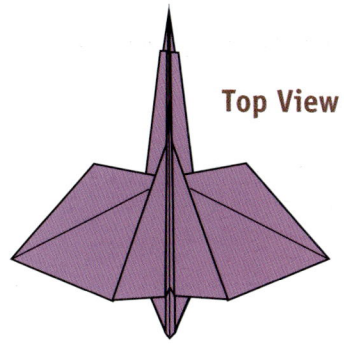

Top View

Canada geese are known for making enormous migrations every year. But how do they know which way they're going, or where they should wind up? Indeed, many birds make such migrations, often traveling thousands of miles. What's really interesting is that the ways birds and pilots navigate are not entirely dissimilar.

Birds migrate using a number of environmental cues. Migrations are usually triggered by changes in daylight; birds migrate as winter approaches and the days grow shorter. It's likely that many also use the sun for directional guidance. In addition, through biological metals found in their brains, many birds are able to sense the magnetic field of the Earth. This gives them a good way to figure out which direction to go.

Pilots have a similar way of navigating. Most aircraft have a magnetic compass, which shows the direction the aircraft is flying. If you know which way you're going, and how fast, that's enough to know where you are. This is called dead reckoning, probably because if you use it you're "dead on." Ships at sea have used this method for centuries. It's somewhat less effective for airplanes, as magnetic compasses (or whiskey compasses, as pilots call them) show errors during ascents, descents, and turns. Many aircraft have gyroscopes, which do a bit better, but suffer their own deficiencies.

Just because you know which direction to go doesn't mean you know where to land. Many birds learn these skills from their parents and the other birds around them. They learn to identify landmarks and other cues that show them their migration routes. For example, the pilots of Operation Migration lead whooping cranes on their annual migration using ultra-light aircraft, which are slow enough for the cranes to keep up. The cranes learn the route, and can eventually fly it by themselves. A pilot who flies the same route over and over can certainly learn the way. I can fly to my two most common destinations just by looking out the window—I don't even need a compass for those.

Though lots of us have such milk runs, most pilots want to fly to places they haven't been before. The simplest means of navigation, beyond compass and stopwatch, relies on navigational charts printed by government agencies. These represent the ground as seen from the air, and are extremely accurate. They also show airports along with their elevation, relative length, and communication frequencies. If you plot your course on a chart, and time landmarks as they're supposed to appear, you can fly just about anywhere. This sort of navigation is called pilotage, and all pilots know it, or at least they're supposed to.

In this day and age the navigation equipment available to pilots is incredible in its accuracy and ease of use. Certainly the gold standard is GPS (Global Positioning Satellite) navigation. This depends on a network of satellites in low earth orbit that transmit signals allowing receivers to pinpoint their position, often within feet. Developed for the military, the GPS array can now be used by anyone. Inexpensive GPS receivers will not only tell you where you are, but, through encyclopedic databases, will also show you what else is there, from towns and roads to airports. Often, these will carry all the relevant data one needs to land at an unfamiliar airport, and are thus phenomenally useful. Between the advent of GPS navigation and new materials in aircraft construction, we are truly entering a second golden age of general aviation.

Not Entirely Abstract Kingfisher

Another variation of the Heron, only a little more so. Begin with the Bird-base Fighter, step 5.

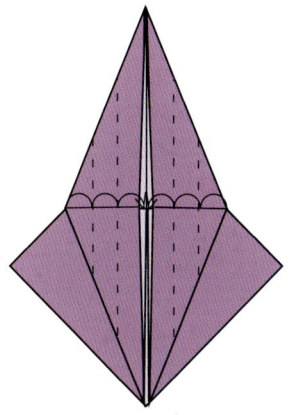

1. Divide the front flaps in thirds, and valley fold them to the middle.

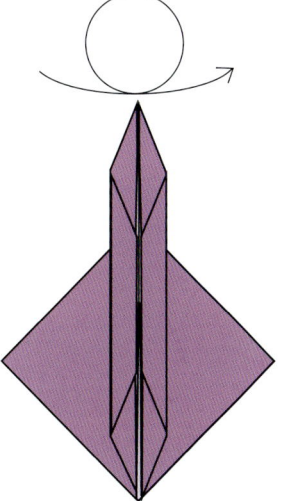

2. Turn over.

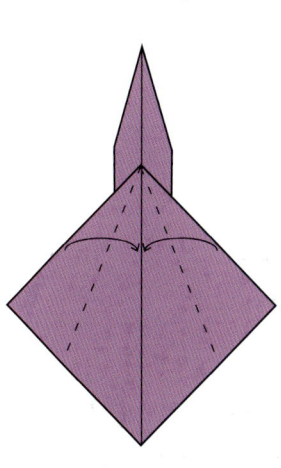

3. Valley fold along the angle bisectors.

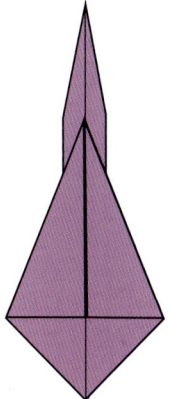

4. Turn over.

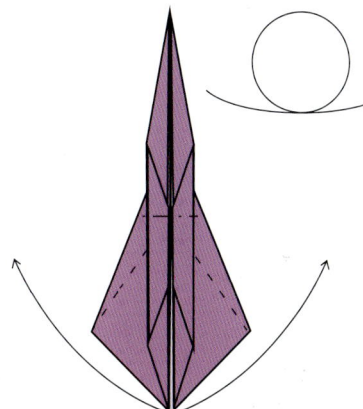

5. Swivel out the wings.

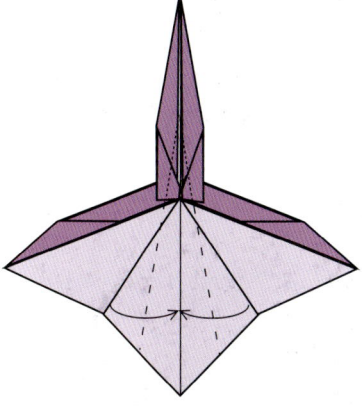

6. Valley fold to make the aft thinner. You'll have to pull the flaps out from underneath the layer above.

Brancusi Animals

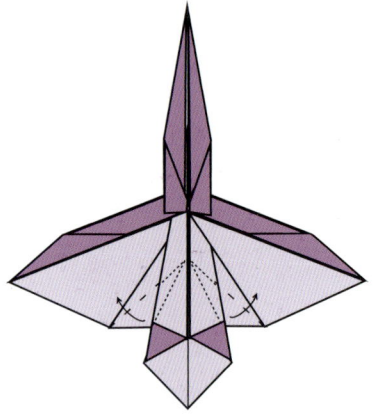

7. Squash fold some paper up from below. The raw edge of the flap folded up should lie just past the folded layer that will come to lie beneath it.

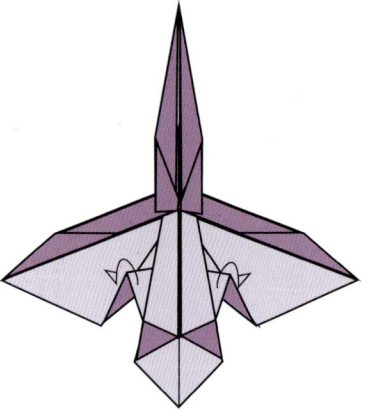

8. Put the top flap underneath the lower layer.

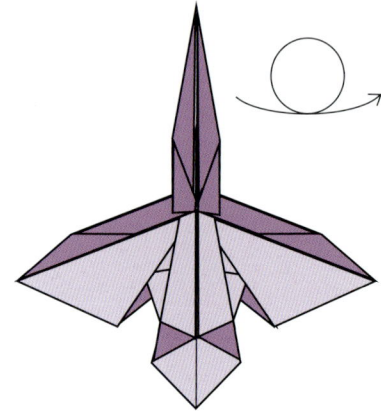

9. Turn over.

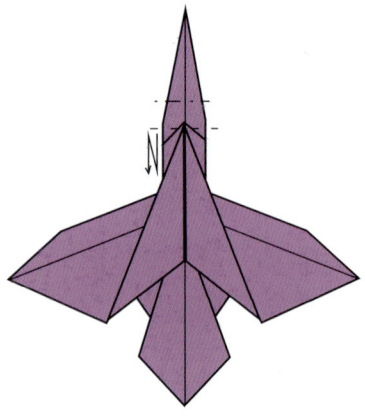

9. Pleat the front and stuff it inside the flap below.

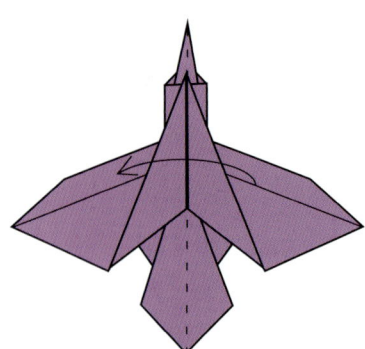

10. Valley fold in half.

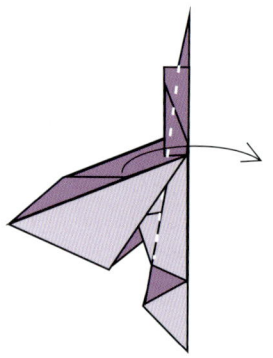

11. Valley fold the wings.

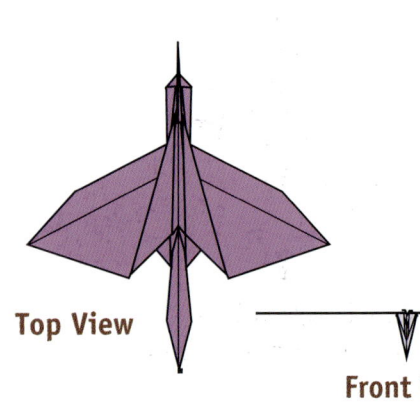

Top View

Front View

Side View

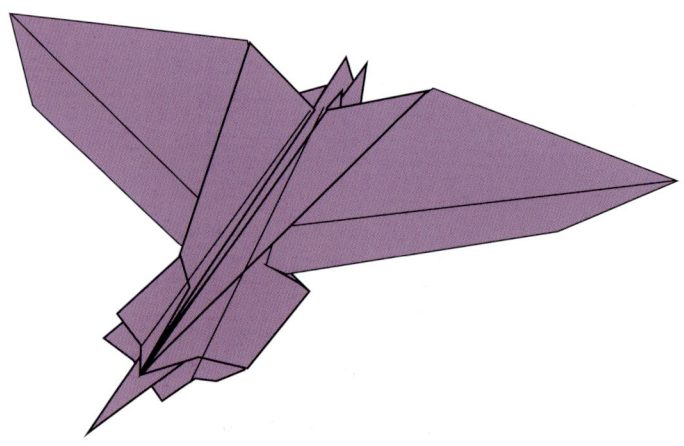

94

Owl in Space

It is called Owl in Space because it really flies (all right, still a bit abstract). Begin with step 5 of the Bird-base Fighter.

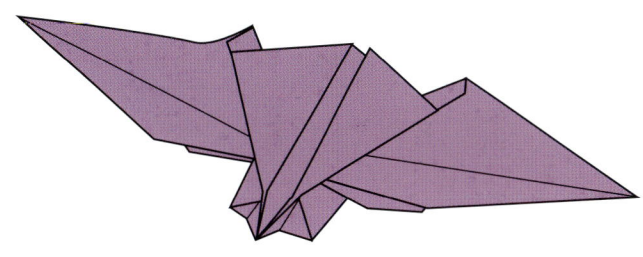

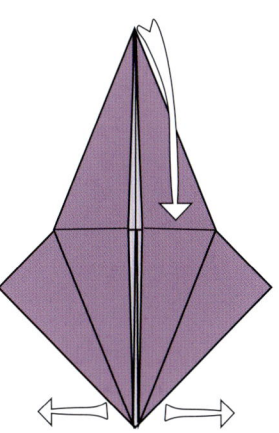

1. Unfold completely.

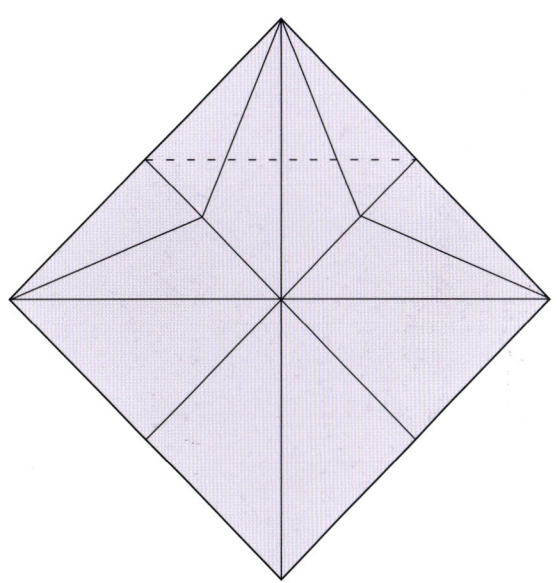

2. Valley fold the top corner into the center.

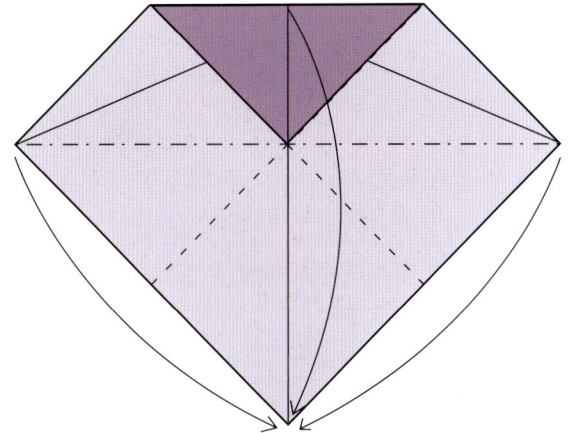

3. Re-form a preliminary base, albeit with the corner inside.

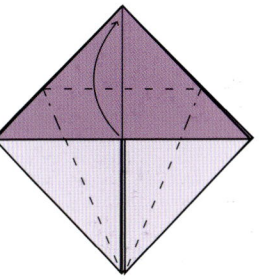

4. Petal fold upward, but only do so on the top layer.

95

Brancusi Animals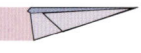

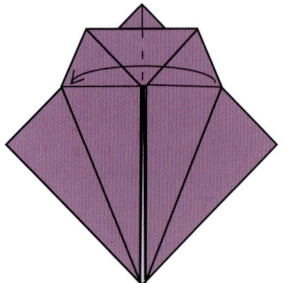

5. Valley fold one flap over to the left.

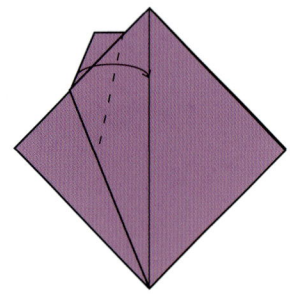

6. Valley fold the flap so that its outside edge lies on the centerline.

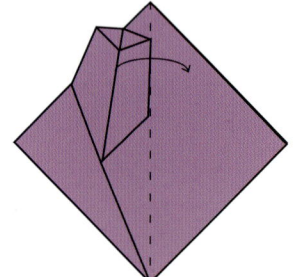

7. Valley fold back over.

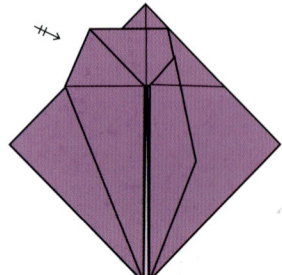

8. Repeat steps 5–7 on the other side.

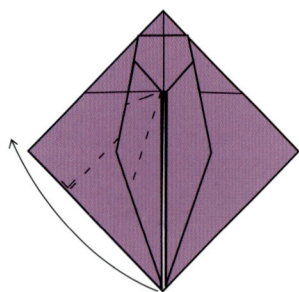

9. Swivel upward to form the wings. Note that the rear of the wings will lie parallel to the raw outer edge of the model.

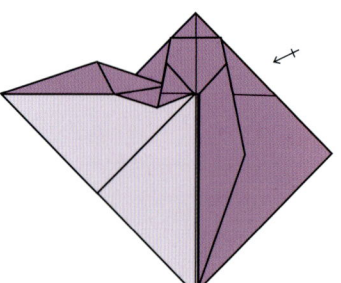

10. Repeat on the other side.

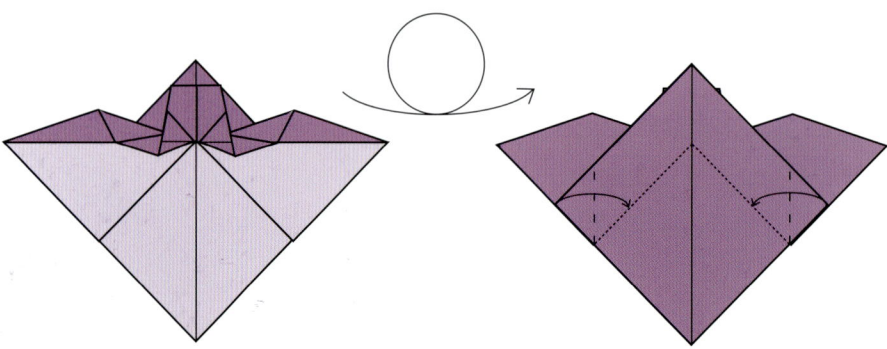

11. Turn over.

12. Valley fold the outer points to the edges of the layer underneath. You can visualize this hidden layer by holding your model up to a strong light.

One can never consent to creep when one feels an impulse to soar. — Helen Keller

Owl in Space

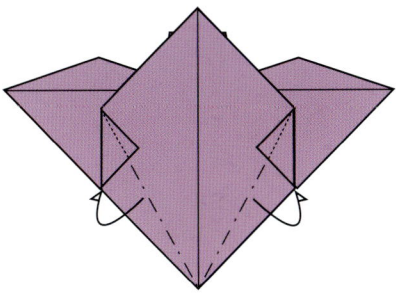

13. Narrow the tail with mountain folds. There will be some extra paper that will swivel behind.

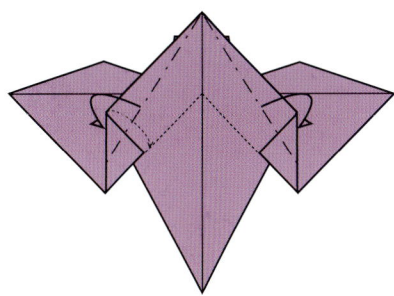

14. Narrow the front with mountain folds. The indicated corner will hit the layer underneath.

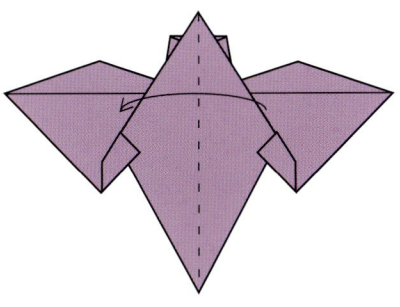

15. Valley fold in half.

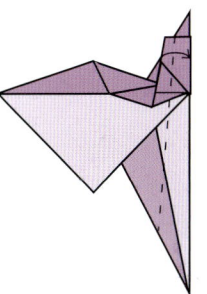

16. Valley fold the wings by lining up the edges shown.

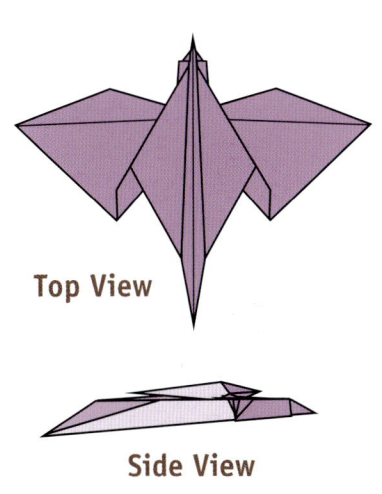

Top View

Side View

Front View

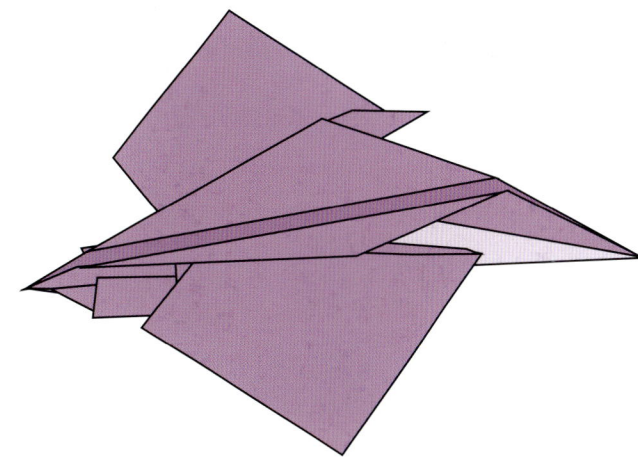

97

Loops, Tubes, and Assorted Mayhem

We are used to seeing aircraft that fit into a mold, machines with a fuselage, wings, and tail, but many aircraft have come and gone that defy this paradigm, and none more so than the flying wings of Jack Northrop.

Northrop was born in Newark, New Jersey, in 1895. His family moved to the West Coast in 1914. In high school he developed a deep interest in aviation, and went to work for the Loughead Brothers (later Lockheed) as a draftsman. In the same year, he moved to Douglas Aviation where despite his extensive education he worked as a draftsman, project engineer, and aircraft designer. In 1927 he returned to Lockheed, where he and Gerrard Vultee designed the Lockheed Vega.

The Vega was an enormously influential design in its day, and showcased Northrop's visionary genius. Pilots were attracted to its rugged design and long-distance capabilities. Amelia Earhart flew across the Atlantic in one. Wiley Post flew a Vega around the world.

The Lockheed Vega

Despite his success designing aircraft for Lockheed, Northrop wanted to pursue his own ideas on the production of all-metal aircraft, a rarity in the late 1920s. He founded his own eponymous company, and built the Northrop Alpha, another revolutionary low-wing monoplane design that introduced such advancements as stressed aluminum skins and deicing boots. His innovations continued with the Northrop Delta and Gamma, but by 1932 his company had become a subsidiary of Douglas Aviation.

In 1938 Northrop sold his interests in Douglas and once again set off on his own. He formed Northrop Aviation, serving as its president and chief engineer. The company produced a number of successful designs for the military, but Northrop's real interest concerned the development of flying-wing aircraft. He felt that flying wings were the future of aviation, owing to their aerodynamically clean properties. But his sleek metal flying wings had propulsion, stability, and production problems, and at a time of shrinking defense spending, the US Air Force abandoned them. As often happens in life, everything old is new again: on November 22, 1988, the air force unveiled the Norton Grumman B2 Spirit Stealth Bomber. The design was a flying wing. Advanced computer controls overcame the wing's stability problems, and other advanced technologies solved its propulsion difficulties. The flying wing gives the weapon one additional advantage for the modern age: the design contributes to a small radar signature, making the aircraft nearly invisible to radar.

Grumman B2 Stealth bomber

Boxoid

I once published an aircraft that was basically a box with wings. I liked it, and made another from a square. Okay, not so original. On the other hand, this one is easy and flies well.

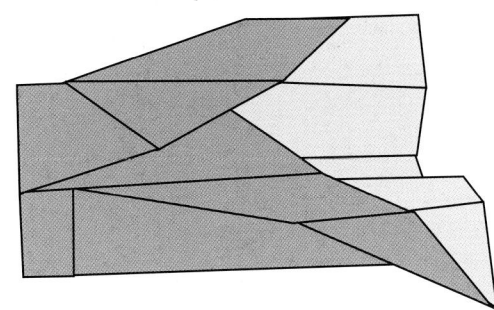

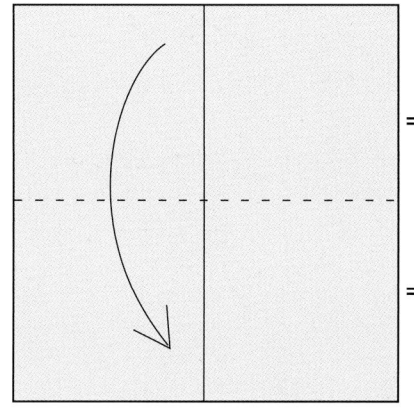

1. Valley fold in half.

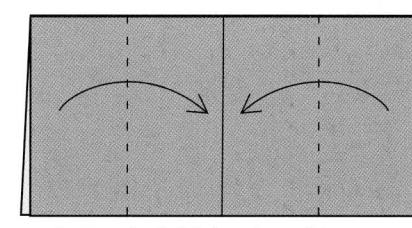

2. Book fold in the sides.

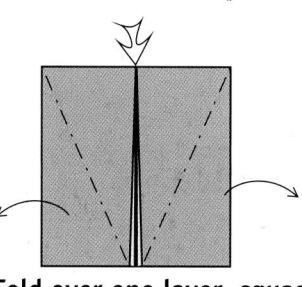

3. Fold over one layer, squash folding the corner. The squash should go all the way down to the outside corner.

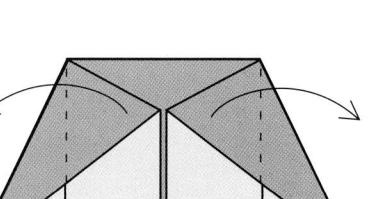

4. Un-book fold (open up with valley folds) on both sides.

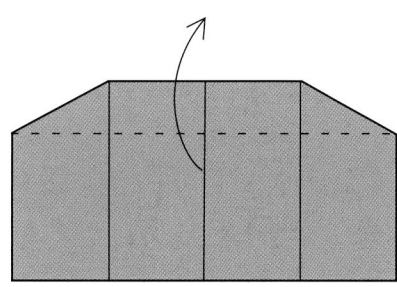

5. Valley fold the front layer upward.

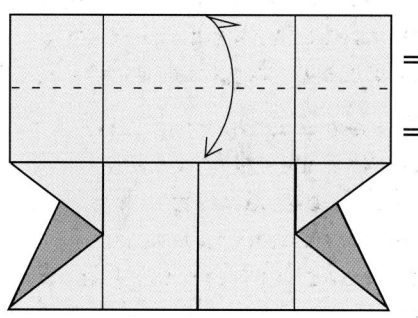

6. Crease the upper flap in half.

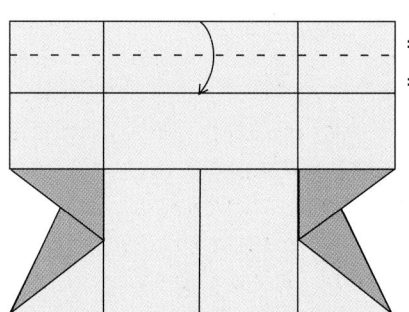

7. Valley fold the top raw edge to the crease made in the previous step.

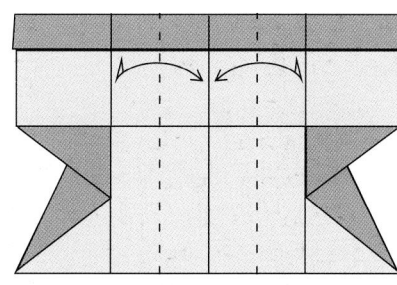

8. Valley fold so that the base of the wings (such as they are) meet the centerline. Crease and unfold.

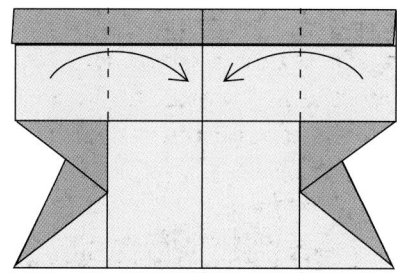

9. Valley fold the two flaps into the middle along preexisting creases.

Loops, Tubes, and Assorted Mayhem

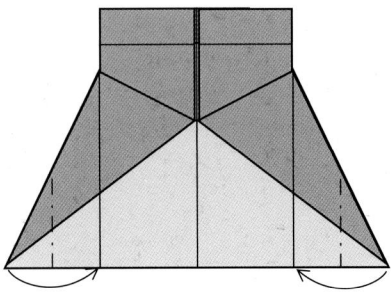

10. Mountain fold the vertical stabilizers so the bottom tips of the paper meet the crease shown.

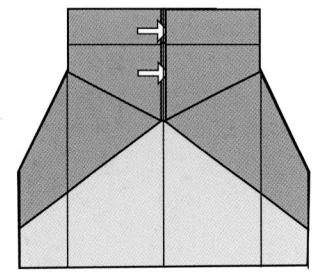

11. Make the aircraft three-dimensional by sliding one of the front corners inside the front flap adjacent.

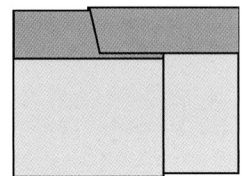

12. A detail of step 11 in progress showing the interior. Slide the two flaps together.

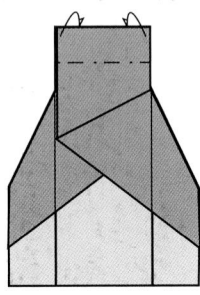

13. Roll the front end inside along the crease already there. This is a bit of a pain, but will firmly lock the front end together.

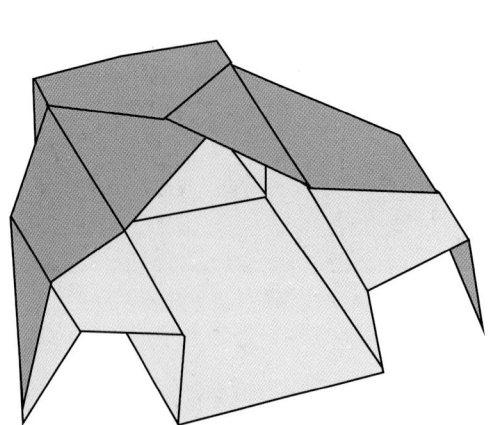

Front View

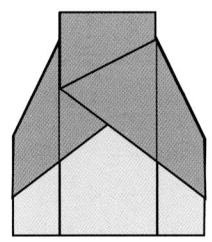

Top View

Side View

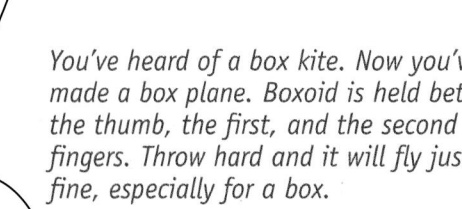

You've heard of a box kite. Now you've made a box plane. Boxoid is held between the thumb, the first, and the second fingers. Throw hard and it will fly just fine, especially for a box.

100

Triangulon

Inherently strong, triangles are found in buildings, bridges, helicopters, and more. They work well in paper airplanes, too.

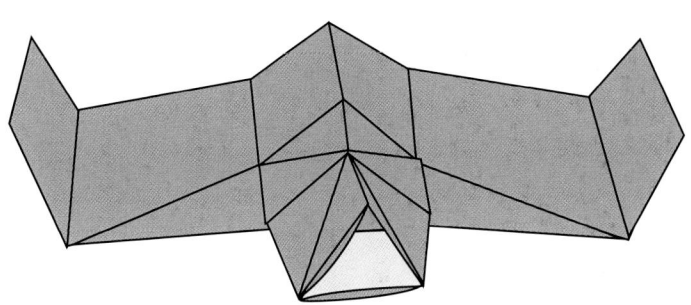

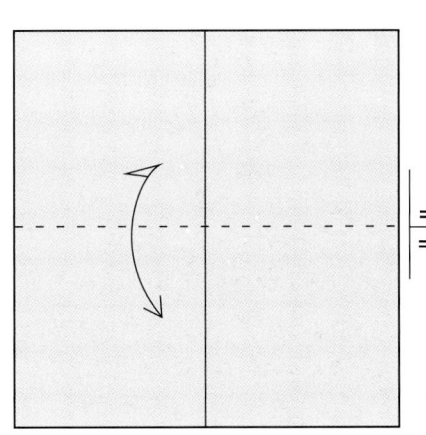

1. Valley fold in half. Crease well, and unfold.

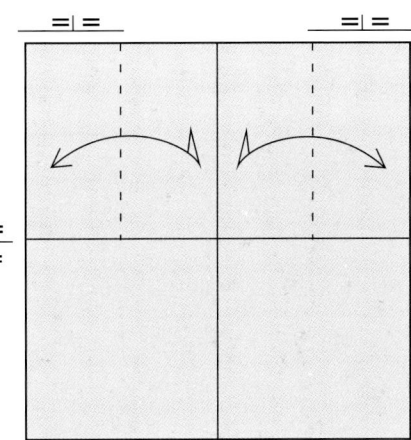

2. Valley fold in half, but only halfway down. Crease well, and unfold.

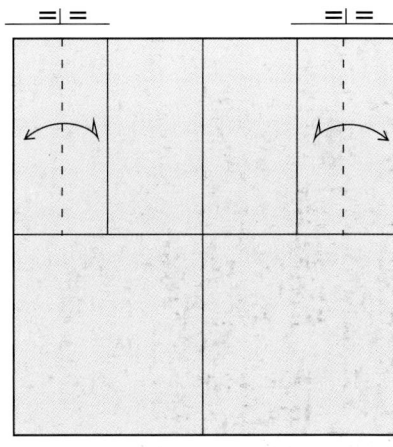

3. Make halfway creases on the outer panel again.

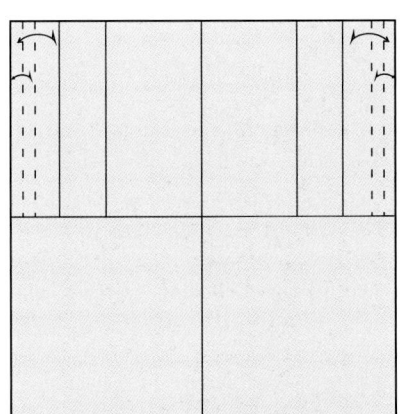

4. Divide the upper panels in sixteenths and thirty-seconds, again only on the upper half.

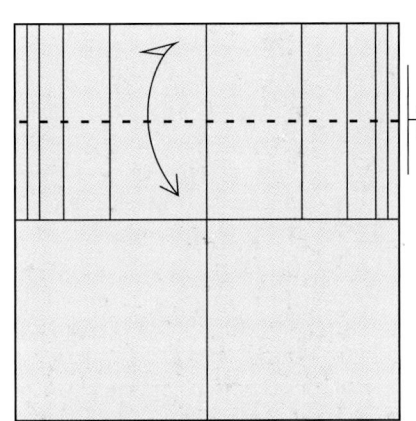

5. Divide the upper panel in half lengthwise.

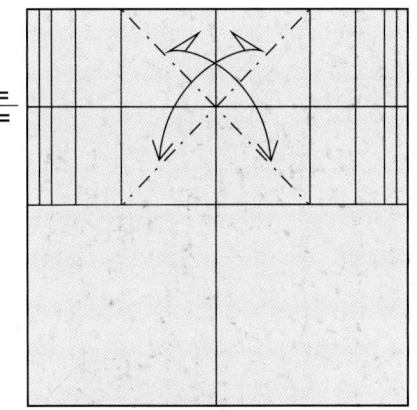

6. Crease the diagonals with mountain folds.

Loops, Tubes, and Assorted Mayhem

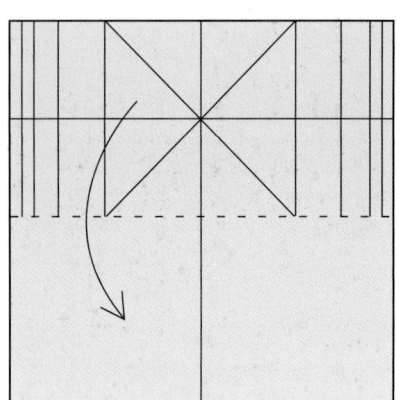

7. Valley fold in half.

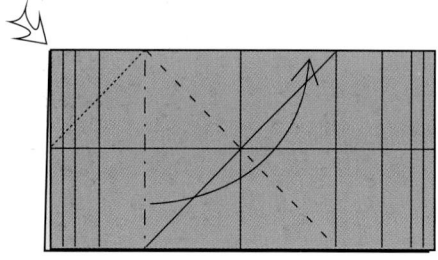

8. Valley fold along one diagonal while mountain folding on one of the vertical creases. The mountain fold will line up with the top, and a valley fold on the lower side will finish the move.

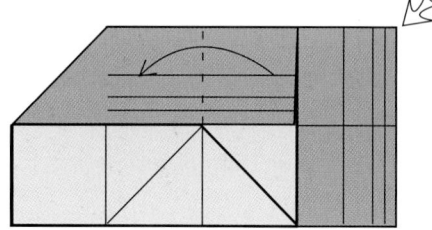

9. Repeat on the other side.

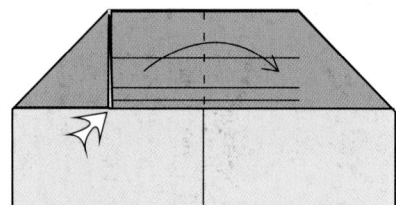

10. Valley fold, squash folding the middle layer.

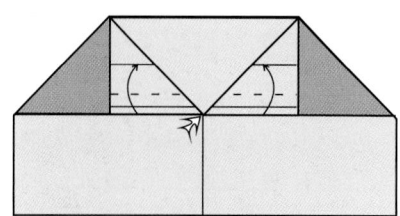

11. Valley fold along a preexisting crease, squash folding the hidden corners.

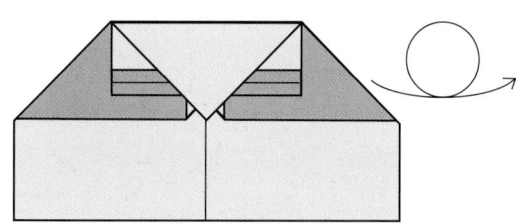

12. Turn over.

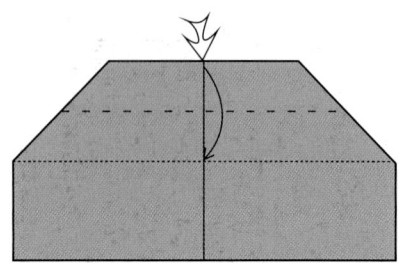

13. Valley fold the top down, squash folding two hidden corners as you do so.

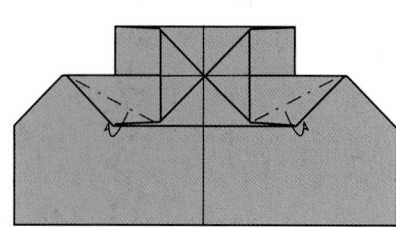

14. Mountain fold the corners behind. Weight forward!

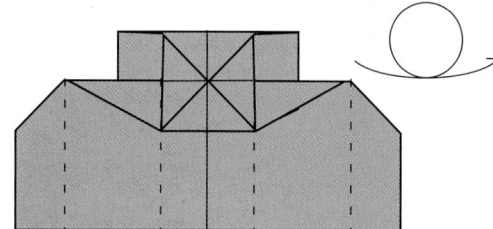

15. Crease valley folds at the indicated spots. These will form the wings and vertical stabilizers. Turn over when you're done.

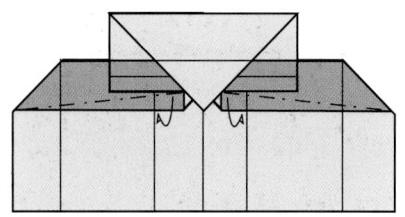

16. Mountain fold the flaps below. It becomes difficult where the flaps are inside.

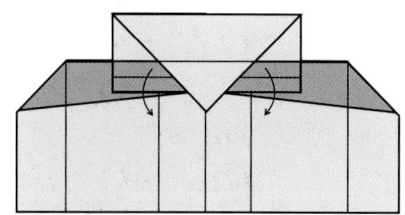

17. Unfold the colored flaps and bring them down.

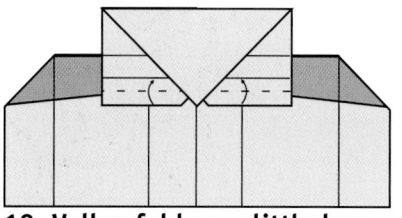

18. Valley fold up a little less paper along a preexisting crease.

Triangulon

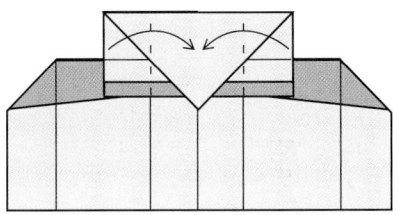

19. Valley fold the front flaps into the center.

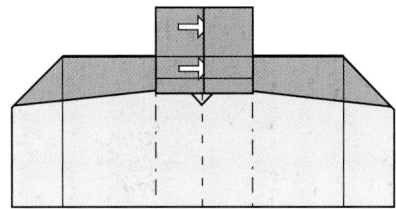

20. Time to put it all together. Move one flap inside the other, while mountain folding the wings. There are two locks.

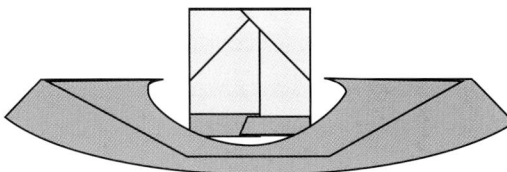

This is an expanded view from above, with the top cut off for your view. The forward corner goes into the adjacent pocket. The lock at the back works similarly to the one used in Boxoid; the corner of one flap goes behind the upper layer of the adjacent flap.

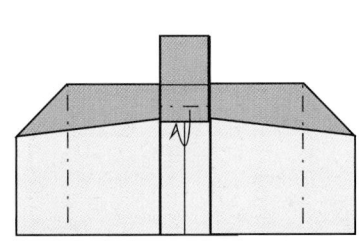

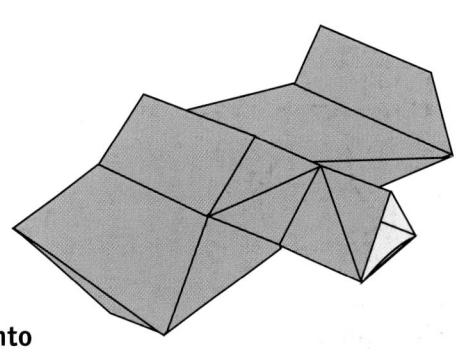

21. Last step, and the most difficult. Roll the bottom into the tube along a preexisting crease. The is the same lock used in Boxoid, but much more difficult to do, because the fold goes in between layers. Fold carefully, and try not to smush anything, as if you do it will cause drag on the airframe. Finish by bringing up the vertical stabilizers.

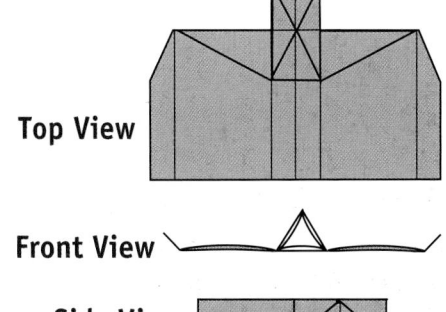

Top View

Front View

Side View

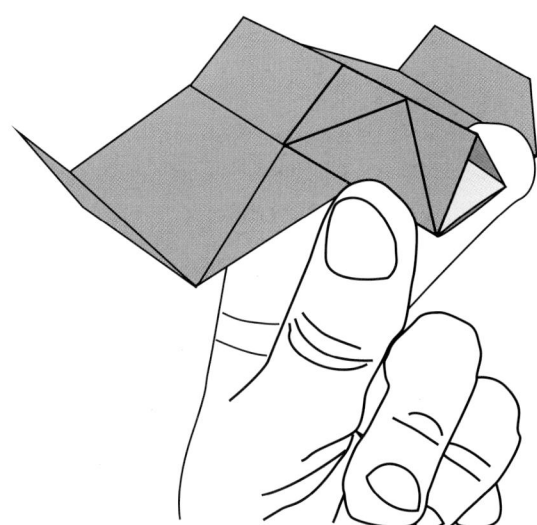

To throw Triangulon, hold the front between your thumb and first finger. Place your second finger behind the front triangular part of the aircraft. Pull it with your fingers when you launch. Throw hard. Also, give it some elevator, and it will fly well.

I like this model, as it looks like an alien spaceship. It doesn't fly as well as some of the other airplanes, because its triangular shape and internal flaps give it quite a bit of drag. Okay, so lots of folding doesn't necessarily make a better-flying airplane, but Triangulon looks neat enough to overcome some of its aerodynamic shortcomings. I hope you agree.

Loops, Tubes, and Assorted Mayhem

Twin Star

The only two-piece model presented herein. Begin with step 14 of Triangulon.

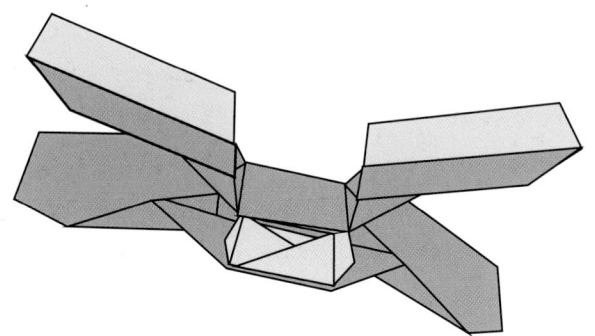

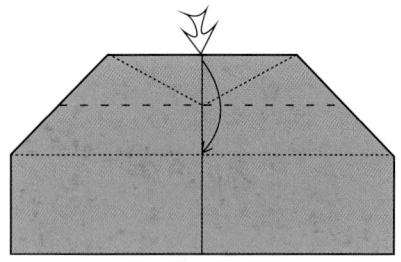

1. Valley fold the top down, squash folding two hidden corners as you do so. The folds in back will lead to the outside corners.

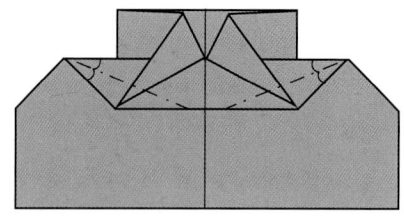

2. Mountain fold the corners behind along their angle bisectors.

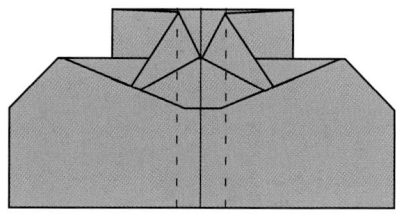

3. Valley fold the wings upward to lie perpendicular to the rest.

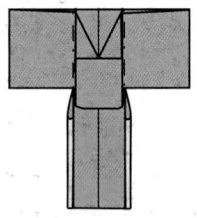

4. Mountain fold the forward flaps down, again so that they lie perpendicular to the body and parallel to the wings.

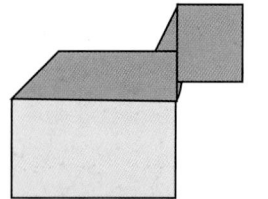

5. To fold a second half, grab a new sheet of paper (same size and color, please) and fold steps 1–4. Fasten the two halves together using their forward flaps (in the same way as fastening the two flaps of the triangulon).

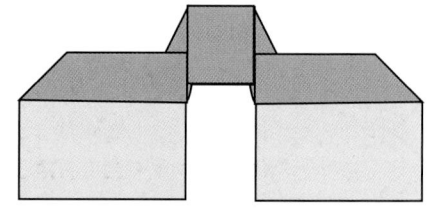

Top View

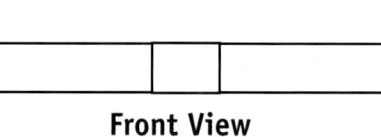

Front View

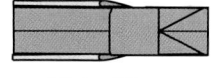

Side View

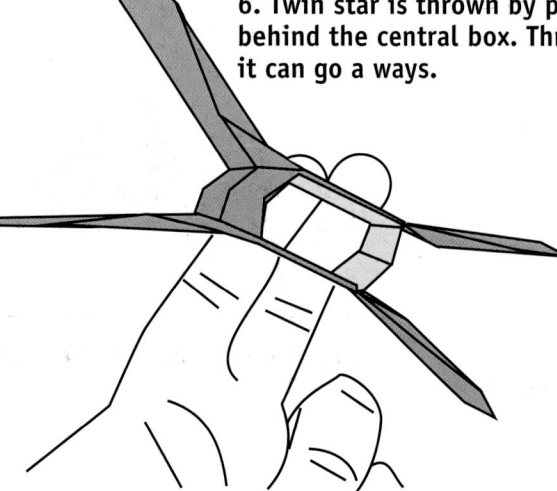

6. Twin star is thrown by placing two fingers behind the central box. Throw as hard as you like, it can go a ways.

Because the two halves of the Twin Star are put on in mirror image, the overall effect is a symmetrical airfoil. In the introduction, you recall, we saw that airfoils generate lift due to their asymmetry. But that's okay, you can see for yourself that a symmetrical airfoil works just fine — the Twin Star can fly. Aircraft built for aerobatics, such as the Pitts biplane, use symmetric airfoils to increase their maneuverability. Though a symmetrical airfoil isn't as efficient as an asymmetrical one, aerobatic airplanes have powerful engines that can easily make up the difference as they dance through the sky.

Nacelle Jet

An airplane with aerodynamic nacelles, where a real aircraft would have its jet engines. Begin with step 5 of the Diamondhead Canard.

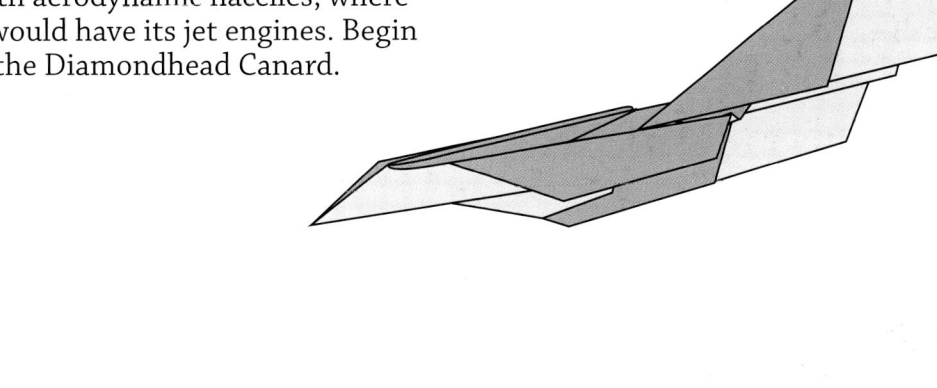

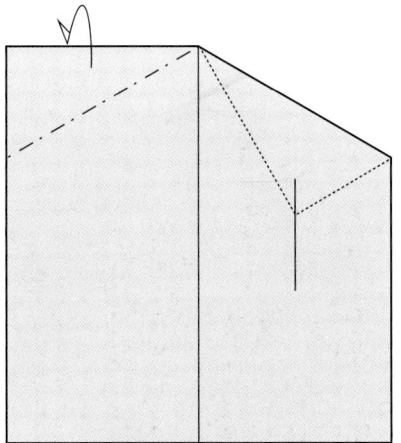

1. Fold the left side to match the right.

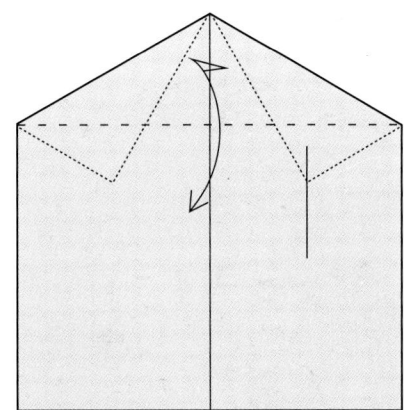

2. Valley fold the top down, crease, and unfold.

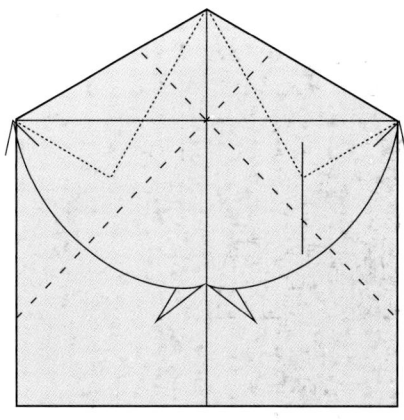

3. Crease two diagonals running through the crease made in step 2, and the centerline.

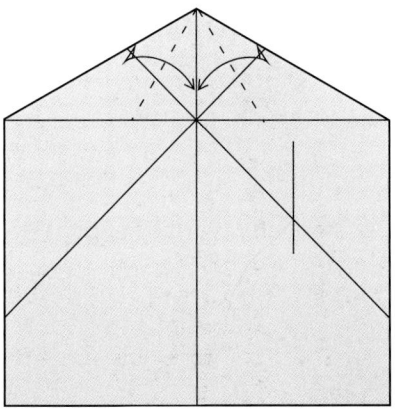

4. Valley fold so that the folded edges at the front lie on the center, but only to the crease made in step 2. Unfold.

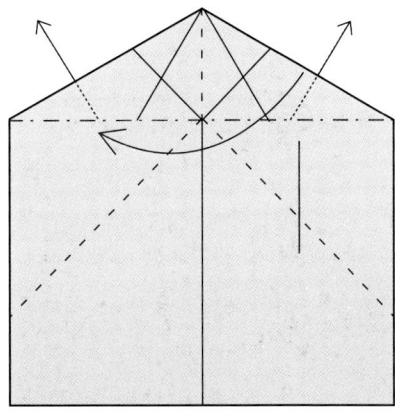

5. Rabbit-ear the front, allowing the underlying flaps to swing out.

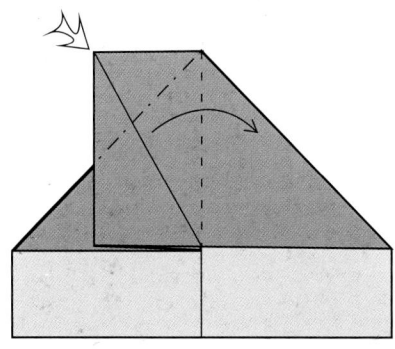

6. Squash fold the resulting flap.

105

Loops, Tubes, and Assorted Mayhem

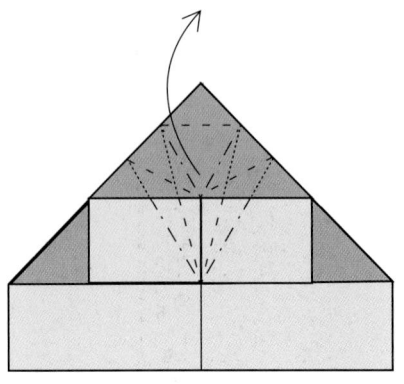

7. Now you get to do a very difficult petal fold. Bring the top layer up along a preexisting crease. Then valley fold the two flaps that you folded at the very beginning of the model.

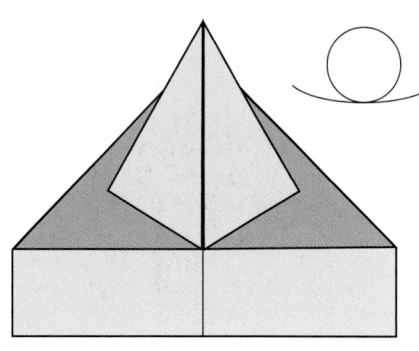

8. Turn over.

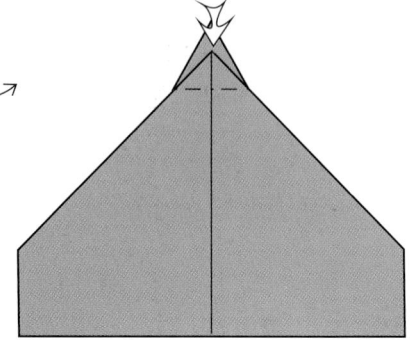

9. Sink the point.

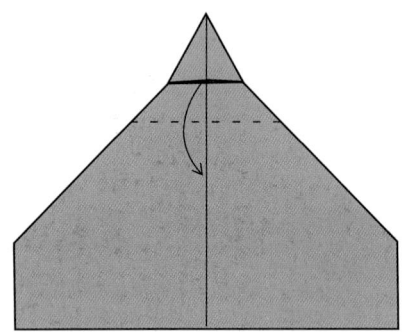

10. Valley fold the top down spreading the sink.

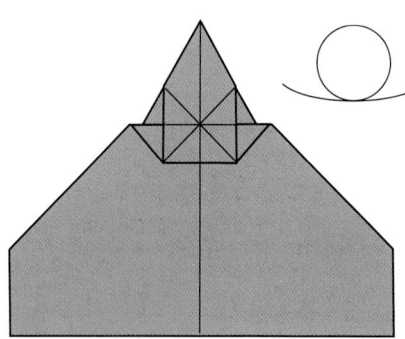

11. Turn over.

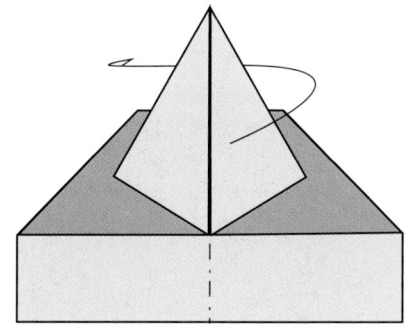

12. Mountain fold in half.

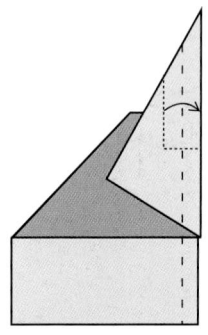

13. Valley fold the wings.

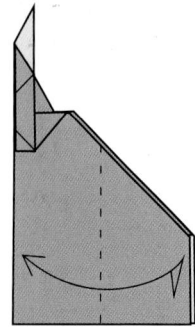

14. Crease the vertical stabilizers.

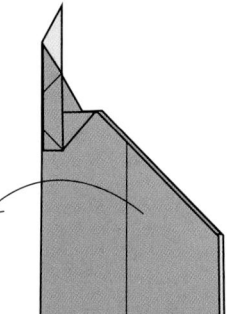

15. Fold the wing back, leaving the flap in front.

There is no excuse for an airplane unless it will fly fast! — Roscoe Turner

Nacelle Jet

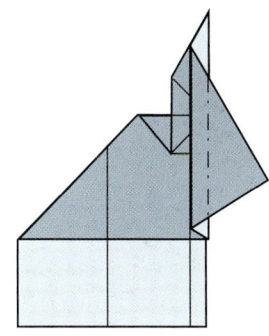

16. Mountain fold the flap where it hits the bottom of the fuselage. Crease and unfold.

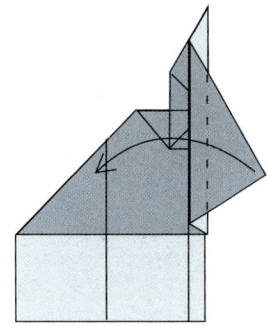

17. Fold back the forward flap.

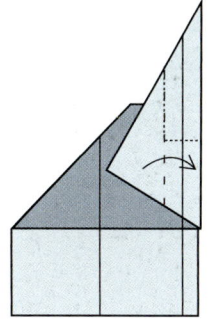

18. Valley fold the flap where it hits the layer below.

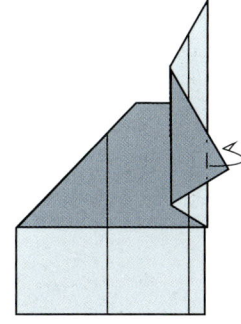

19. Tuck the end of the flap into the pocket below it.

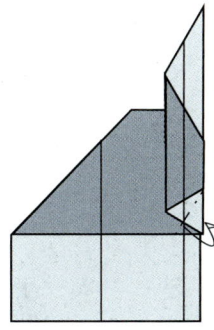

20. Mountain fold a bit of paper behind to lock in the nacelle. The fold should originate at the back of the nacelle, and run through the intersection shown.

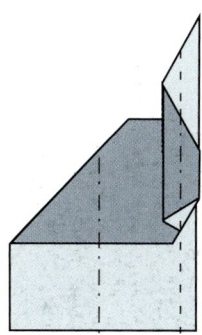

21. Fold the wings and stabilizers, and open out the nacelles.

Given an easy throw, the Nacelle Jet will give good glides. But why does a jet have nacelles, anyway?

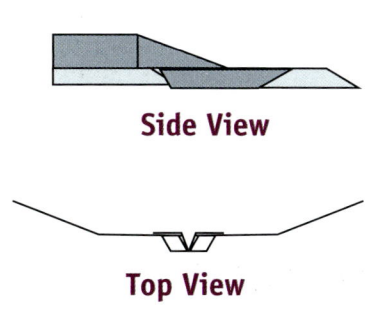

Side View

Top View

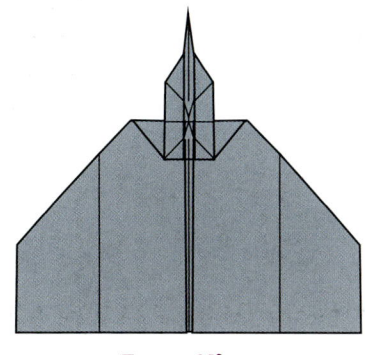

Front View

Loops, Tubes, and Assorted Mayhem

Aircraft require a great deal of force to become and stay airborne; thus aircraft engines are very powerful. They all burn fossil fuels derived from petroleum, and work on similar principles. They vary in the particulars, and include piston, turbine, and turbofan engines. Rocket engines are used for aircraft that venture into the icy reaches of space.

Piston engines are so named because a piston moves up and down inside a combustion chamber, drawing in air and fuel, compressing them for burning, then expelling the exhaust gasses. All this activity spins a crankshaft, which is attached to a propeller in airplanes. Most of the engines you encounter in your everyday life work similarly.

Turbochargers and superchargers can increase the power output of any engine by pushing more air into the cylinders. More air can be burned by more fuel, thus increasing engine power. Both work using compressors: superchargers have an engine driver compressor, while turbochargers use the exhaust gasses to run their compressor.

Some airplanes have turbo-normalized piston engines, whose exhaust-driven compressor maintains normal air pressure in the engine. These are used to get the aircraft into the stratosphere, where the air is too thin for normal piston engines to function. The advantages to this are that one often gets huge tailwinds at such heights; also, most weather is below the stratosphere, allowing one to fly over ice and storms that would stop other light airplanes. And it's easier for wings to move through such thin air. The downside is that the air at such altitudes is too thin for humans as well as engines; thus the aircraft must be pressurized, or the pilots and passengers must use supplementary oxygen. Examples of both are quite common.

A piston banging in and out is loud and tends to produce vibration. Because of this most aircraft engines must be overhauled with a given frequency, usually designated by the manufacturer. In addition, there are performance limits, since propellers can spin only so fast. Eventually the tips of the propeller blades reach the speed of sound, at which point drag and other factors diminish their efficiency. So at high speeds and high altitudes, turbine engines reign.

A turbine engine doesn't have a piston banging around, but is arranged around a central shaft.

Suction   Compression   Combustion   Exhaust

crankshaft

Most piston engines function in four cycles. In the first, air and fuel (blue) are pulled into the combustion chamber by suction. They are then compressed. A spark from the spark plug ignites the mixture, whose explosion powers the descent of the piston (black arrow). The exhaust (red) is then pushed out.

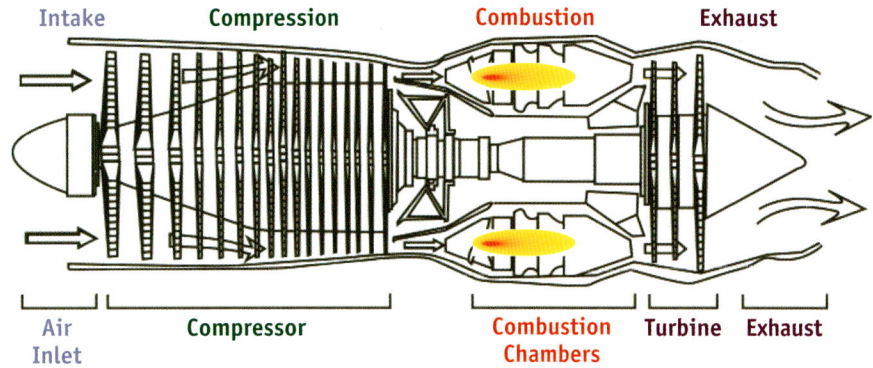

A schematic of a turbine engine. Notice that all the things that happen sequentially in a piston engine (compression, combustion, exhaust etc.) happen simultaneously in a turbine engine.

Air enters from a large opening at the front, and goes through a series of compressors before mixing with fuel and being ignited in the combustion chambers. The hot expanding gasses run a turbine, which powers the compressors. The hot gasses then leave the exhaust at the back.

Many airplanes and helicopters have turboprop engines, where the majority of the engine's power drives the central shaft that's attached to the propeller. Though they suffer some of the limitations of piston aircraft (the propeller tips can't travel faster than the speed of sound, for example) they can easily operate at high altitudes. Also, turbines have much less vibration than piston engines, and are thus quieter and have little wear and tear. The overhaul frequency of a turbine engine is much less than that of a piston engine, making them more economical to operate. That's why most military and airline travel is turbine powered.

A turbofan engine operates on the same principle as a turbine engine, but derives most of its motive power from the exhaust. The engine emits exhaust gasses backward at high velocity; the reaction of the aircraft, governed by Newtonian physics, is to move forward similarly. Turbofans allow a great deal of air to bypass the compressors, mixing it with the hot gasses to increase the volume of the exhaust and the power of the engine. Turbofan engines are found on jets, and operate at high altitudes and speeds.

To move truly high and fast, however, one needs a rocket. These act much like jet engines, exhausting hot gasses at high velocities. But rockets carry their own oxidizer, allowing them to traverse outer space, where a jet engine could not function due to the lack of air. Rockets can have solid or liquid fuel, or a mixture of the two. The Saturn V, the largest and most powerful rocket ever built, used liquid oxygen as its oxidizer, and liquid hydrogen for fuel.

A rocket engine is really a jet engine that carries its own oxidizer. Most engines use atmospheric oxygen to burn their fuel. Rocket engines carry a chemical oxidizer that's mixed with fuel in the combustion chamber. The resulting gasses are expelled from a jet nozzle, propelling the craft forward.

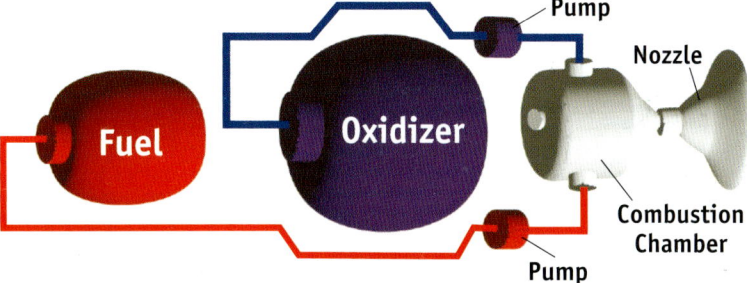

Stands

Rocket flights date back to the invention of gunpowder in Imperial China 1,000 years ago. But gunpowder has never been sufficient to lift a vehicle into space, and no rocket could do so until Robert H. Goddard begin experimenting on liquid-fueled rocket engines.

Goddard was born in Massachusetts in 1882. Inspired by H. G. Wells' *War of the Worlds,* he became enamored of rocketry at an early age (on October 19, 1899, he claimed, and celebrated this "anniversary" all his life). Goddard received his Bachelor of Science degree from the Worcester Polytechnic Institute, and his master's and doctorate from Clark University. He took on a fellowship position at Princeton in 1912, though he had to leave in 1913 due to a bout with tuberculosis.

It was during his recovery that he pursued his most influential work on rockets, with a patent describing multistage rockets, and another describing a liquid-fueled rocket engine powered by gasoline and liquid nitrous oxide. In 1919 the Smithsonian published Goddard's pioneering work on rocketry, which heavily influenced later space scientists. His work was met with derision from the press, however, and he became very private about his discoveries. In the 1920s Goddard began experimenting with liquid-fueled rockets and met with success; indeed the site where his first rocket blasted off is a national landmark. Rocket studies were and remain expensive, and Goddard's diminutive professor's salary limited his experiments.

That is, until he was contacted by Charles Lindbergh in 1929. Recognizing the importance of Goddard's work, Lindbergh arranged funding for him. Goddard relocated to Roswell, New Mexico, where he built rockets with increasing sophistication, inventing gyroscopic control mechanisms and more powerful engines. Sadly, upon seeing the remains of a crashed V2 rocket, he realized that his brainchild had been used for great evil, designed no more rockets, and died of cancer in 1945. Goddard's legacy cannot be overstated, as he laid the scientific and ideological groundwork for those who eventually did reach the heavens.

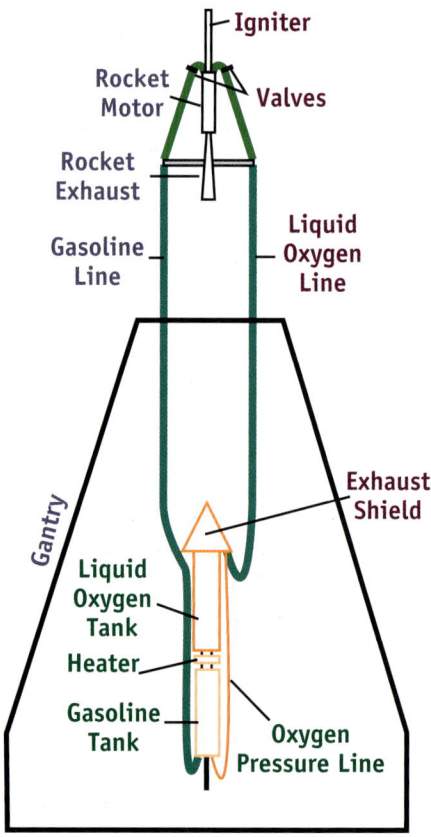

An early liquid-fueled rocket launched by Robert Goddard

Stand Long

A simple stand that begins with a square of paper (the same size you used to fold the airplane) white-side up.

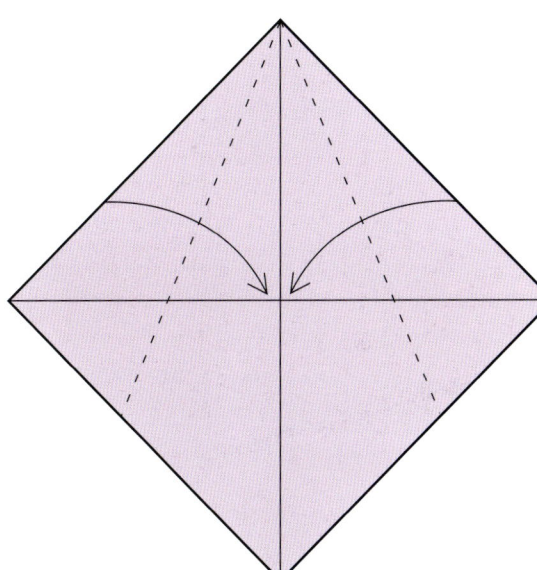

1. Valley fold so that the edges fall on the centerline.

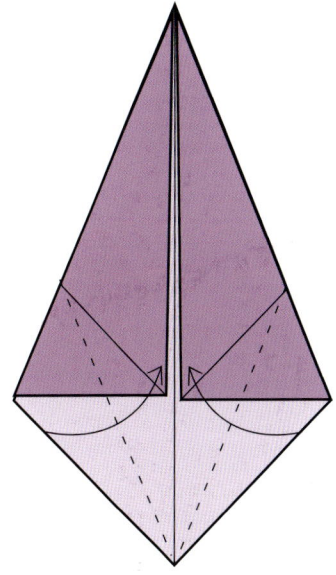

2. Repeat on the bottom to make a nice diamond.

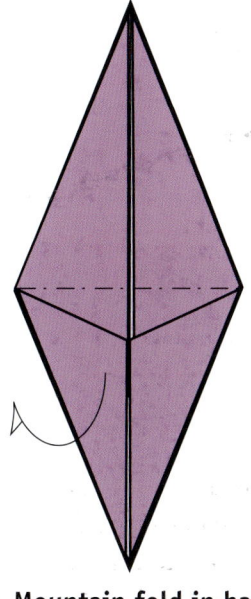

3. Mountain fold in half widthwise.

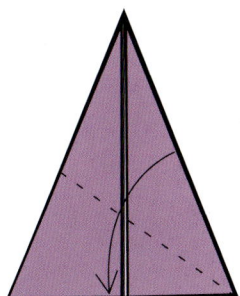

4. Valley fold so that the outside folded edge lies on the bottom.

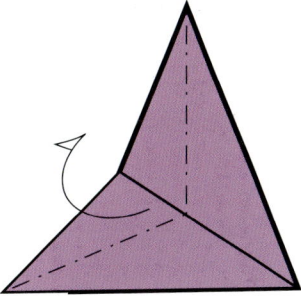

5. Mountain fold in half.

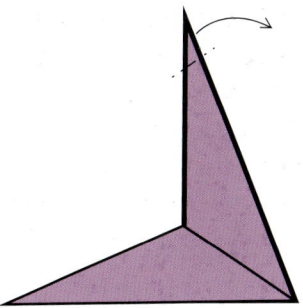

6. Inside reverse fold the tip.

Never fly the 'A' model of anything. — Ed Thompson

Stands

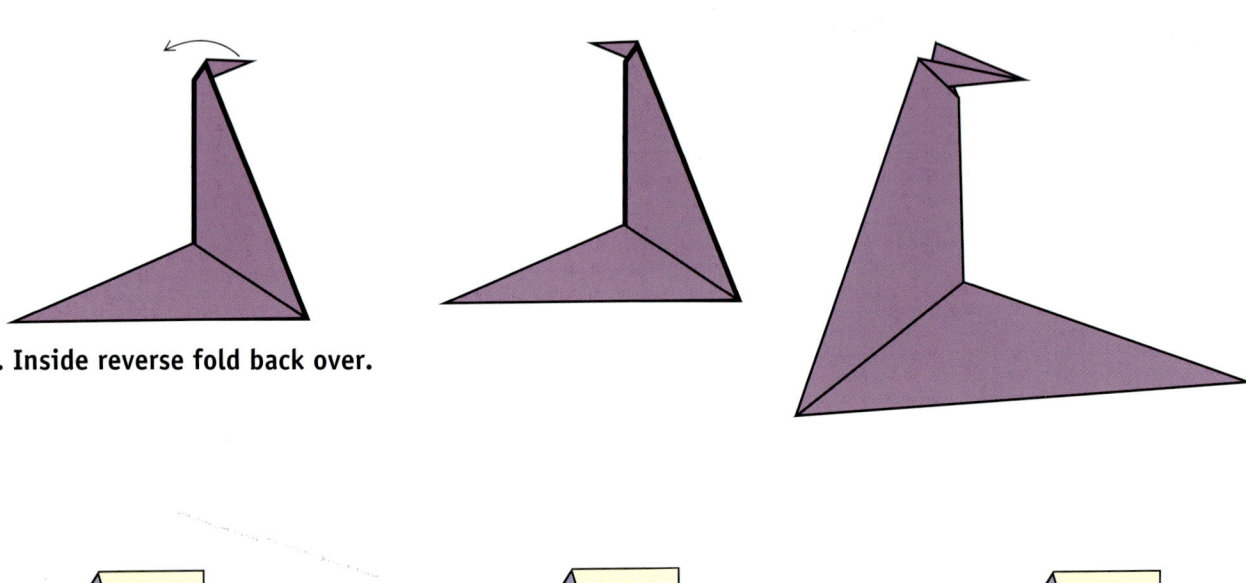

7. Inside reverse fold back over.

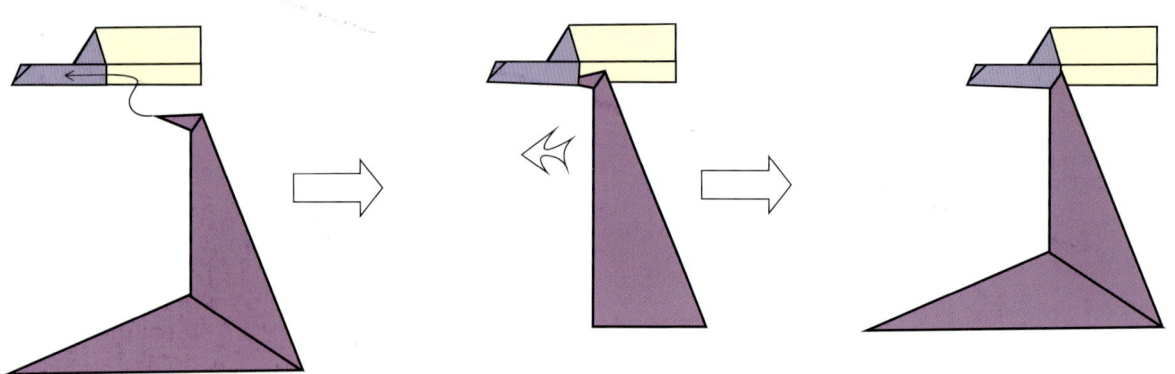

To use this or any of the following stands, find a pocket in the airplane you want to display. For example, Wild One, from the first chapter, has an overlapping layer in the front. This forms a pocket into which you can insert the point of the stand. Once fully inserted, the stand will hold the aircraft quite securely for display, and you'll have a nice little aviation-themed work of art. Don't use fancy patterned papers or foils for your stands; it's the airplane you want to show off. Different stands work better for different airplanes, so I've included several.

> *Up here with the song of the engine and the air whispering on my face, I am completely, vibrantly alive.* — **Stephen Coonts**

Stand Tall

A very effective and easily assembled stand that begins our association with the Fish Base.

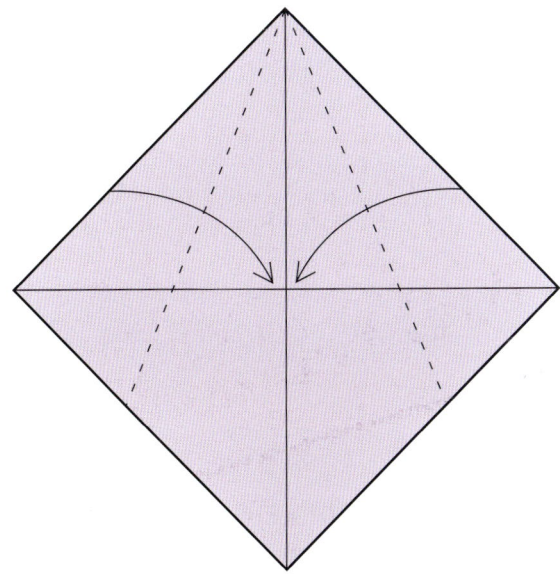

1. Valley fold so that the edges fall on the centerline.

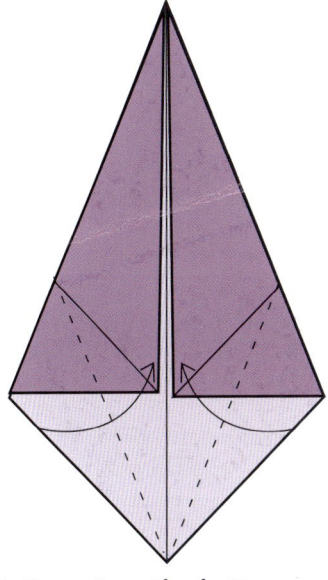

2. Repeat on the bottom.

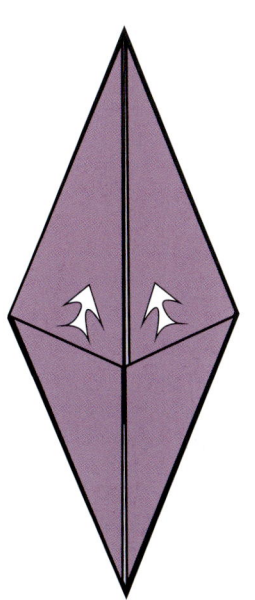

3. Pull out the hidden flaps of paper and flatten.

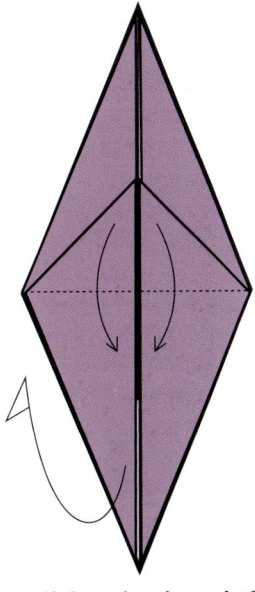

4. In traditional origami this is the Fish Base. The flaps make fins. Stands don't have fins. Mountain fold the lower flap behind, allowing the shorter flaps to flip down.

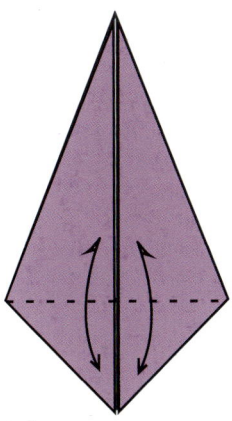

5. Valley fold the short flaps up along their bases, crease, and unfold.

Stands

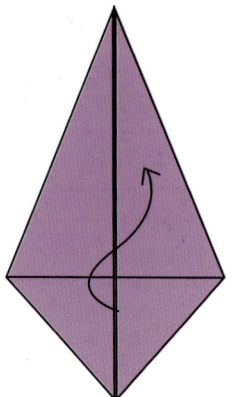 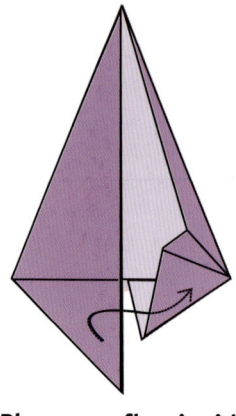 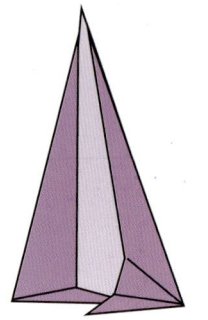 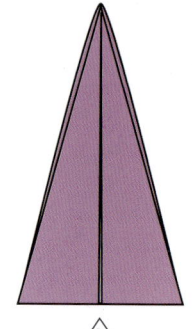

6. Lift up one raw edge and open up one side.

7. Place one flap inside the open flap adjacent to it. Imagine the inside flap is embarrassed.

8. Step 7 in process.

9. Once the adjacent flap is all the way inside, close back up.

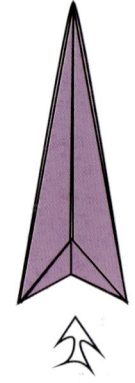 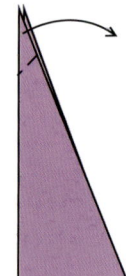 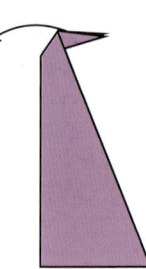 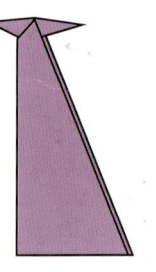 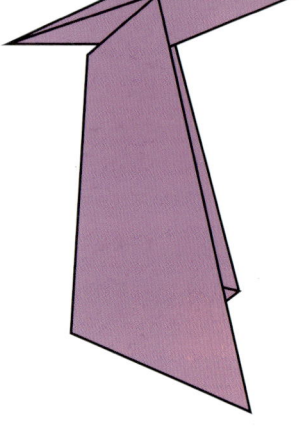

10. Closing in progress.

11. Side view. Reverse fold the two tips together.

12. Reverse fold the inside tip back the other way.

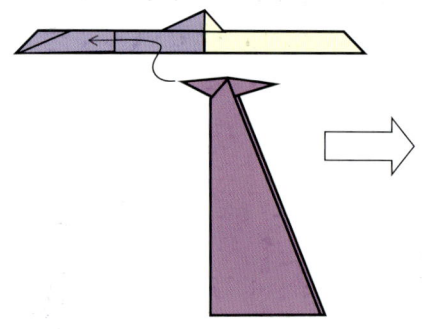 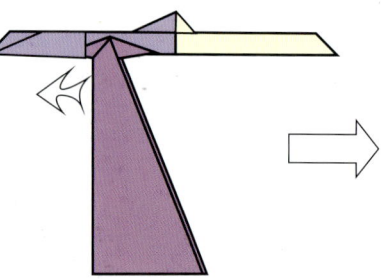 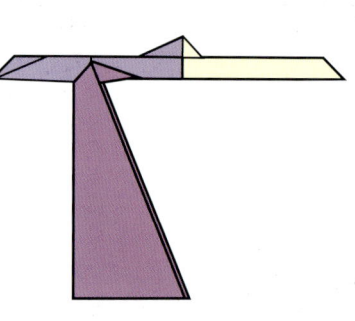

This stand works similarly to the previous one. Here we see that Sharkie, from the first chapter, has an overlapping layer in the front into which you can insert one of the points of the stand. The other holds the airplane up from the rear. I could use someone to do that for me.

Stand Open

Similar to the last two, but with an open base for greater elegance. Start with step 4 of Stand Tall.

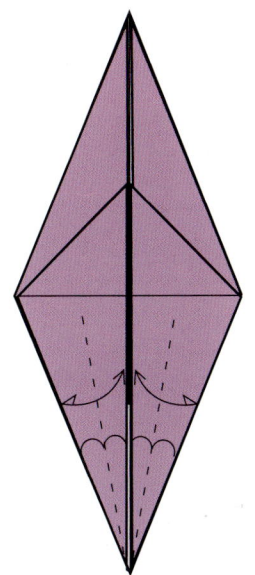

1. Partway crease the angle bisectors lengthwise.

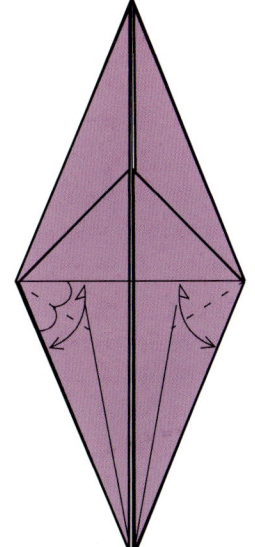

2. Partway crease the angle bisectors widthwise.

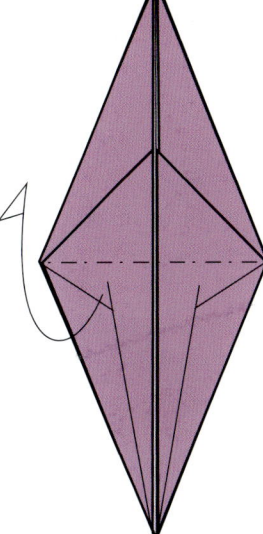

3. Mountain fold the bottom behind.

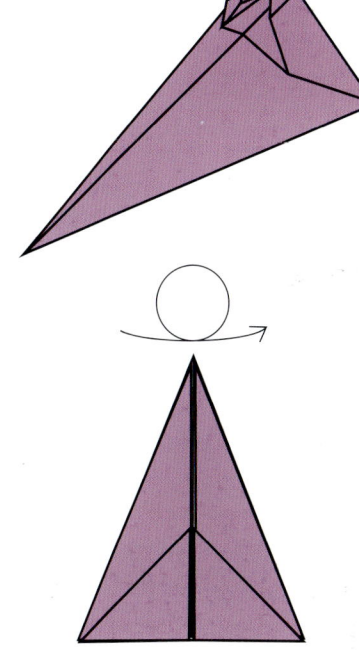

4. Turn over.

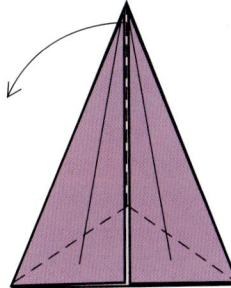

5. Rabbit ear the top flap.

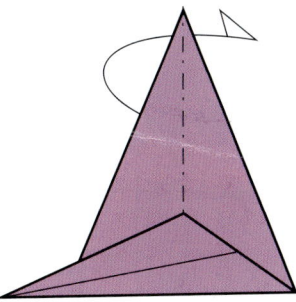

6. Mountain fold the back flap in half.

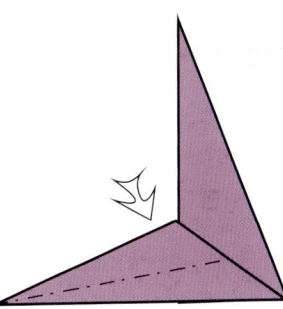

7. Open sink along preexisting creases.

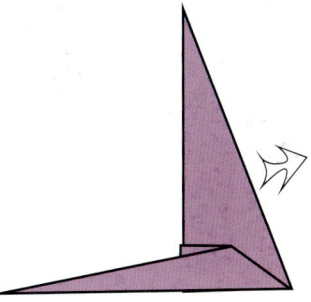

8. Open out.

9. Bottom view. Place one flap inside the open flap adjacent to it, just as in the Stand Tall.

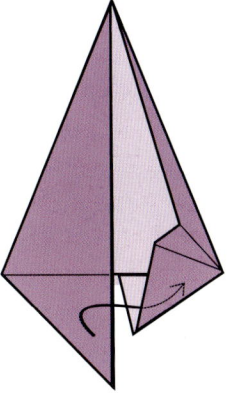

10. Once the adjacent flap is all the way inside, leave it open.

11. Inside reverse fold the tip.

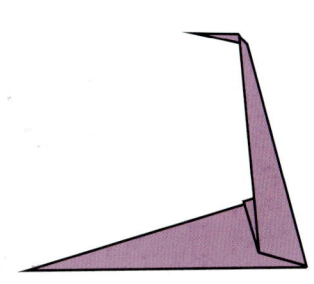

All done! The slimmer point up top works well with airplanes that have slim fronts.

115

Stands

Stand Sharp

The fish base is certainly getting a good work out in this chapter! This stand is as streamlined and swoopy as many of the airplanes. It is a little more difficult to fold, though. Start with step 2 of Stand Open.

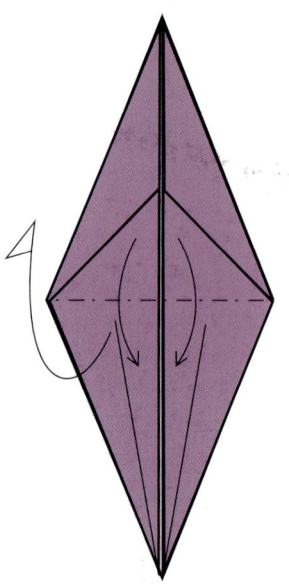

1. Mountain fold the lower flap behind, allowing the shorter flaps to flip down.

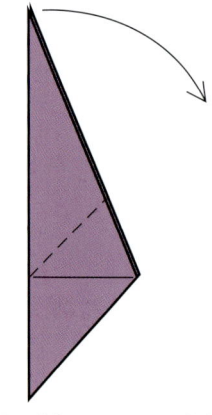

2. Mountain fold in half lengthwise.

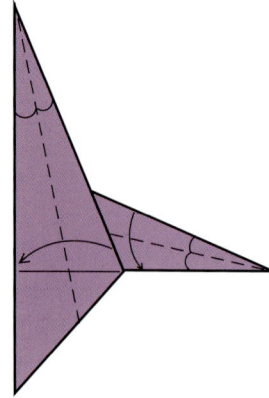

3. Inside reverse fold the inner flap.

4. Valley fold along the angle bisector. This will bring up a pocket, which can be resolved by folding the upright in half.

5. Valley fold the lower part of the flap along an angle bisector.

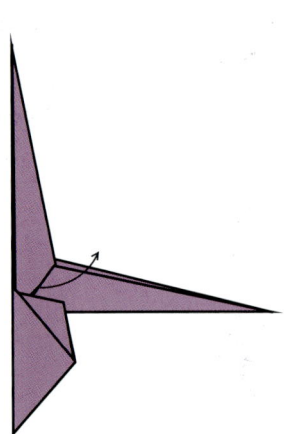

6. Unfold the flap.

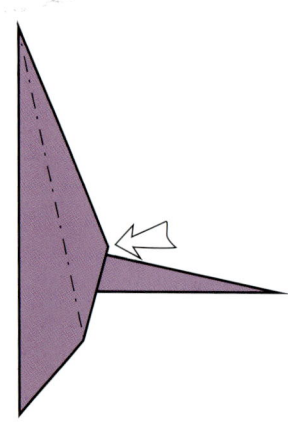

7. Closed sink the flap.

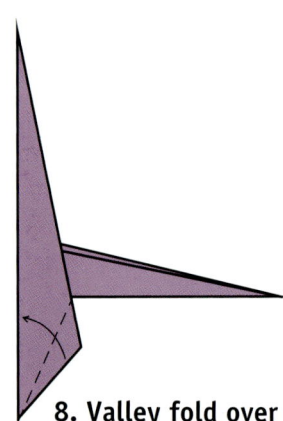

8. Valley fold over as far as you can.

116

Stand Sharp

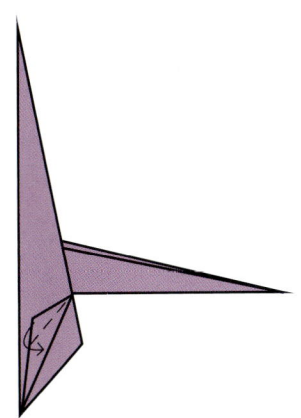

9. Valley fold again.

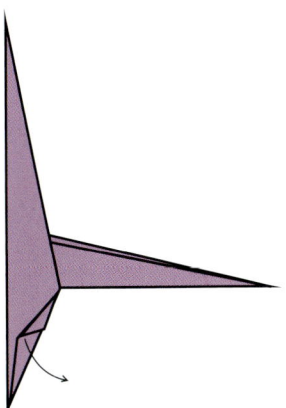

10. Unfold.

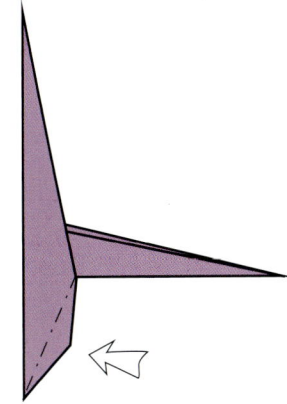

11. Closed sink again. I could have called this the closed sink stand, but it didn't sound as good.

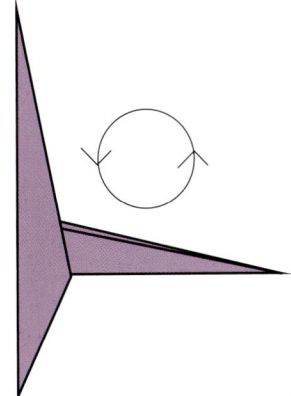

12. Rotate 90°.

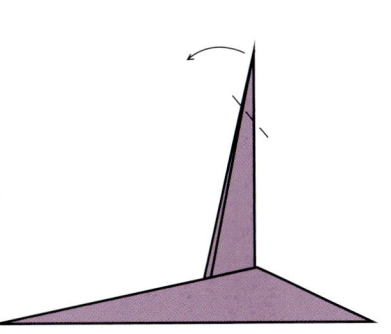

13. Inside reverse fold a little bit off the top to hold your airplane.

14. To finish, valley fold each of the shorter flaps out perpendicular to the rest. They will form the supports for the stand.

Stands

Pylon Stand

Okay, now for something completely different. Some of the stubbier—well, all right—less swoopy aircraft don't look that good on a really swoopy stand. Pylon stand is non-swoopy, sturdy, and can fit its holder into very tight spots.

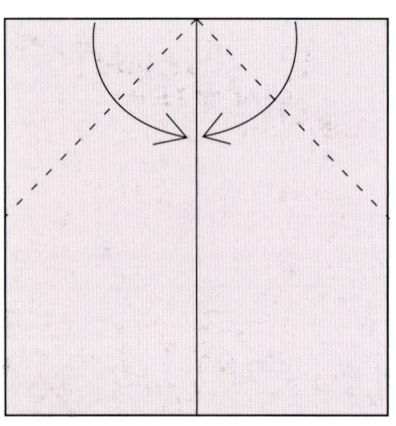

1. Begin with a square creased in half lengthwise. Valley fold the corners so that the top edges lie on the center. Turn over.

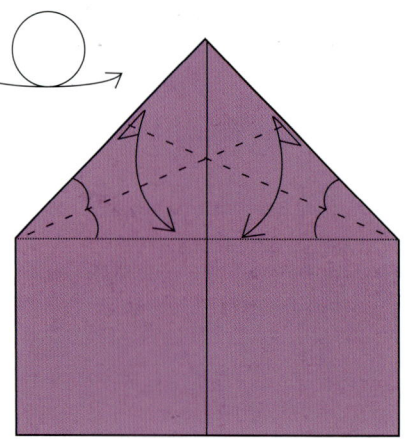

2. Crease the angle bisectors.

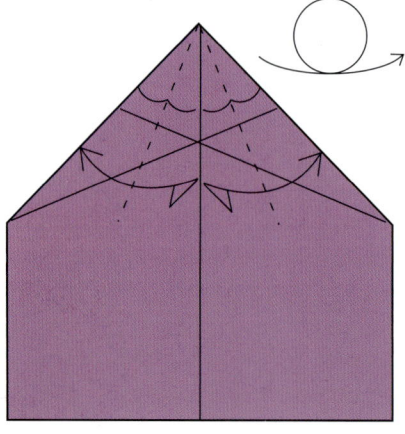

3. Crease more angle bisectors. Turn over.

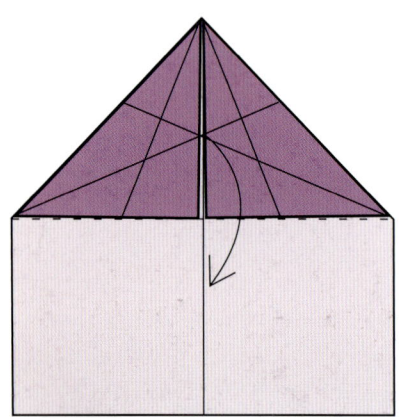

4. Valley fold the colored triangle at the top downward.

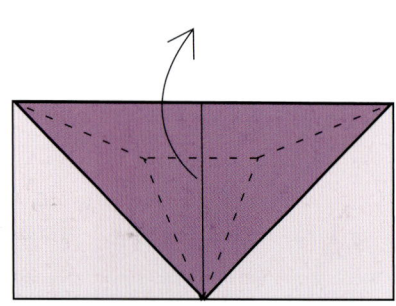

5. Collapse along those bisectors, narrowing the top point and bringing it upward.

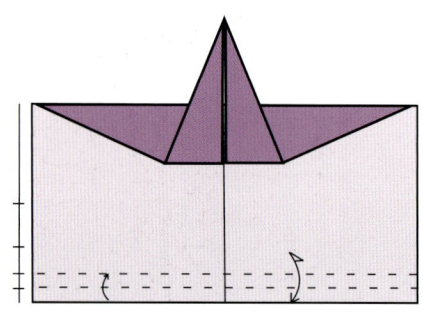

6. Crease along the 1/8 and 1/16 lines, then valley fold along the 1/16 crease.

Pylon Stand

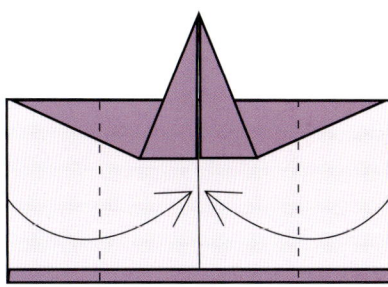

7. Valley fold so that both outside raw edges rest on the midline.

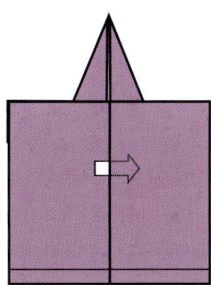

8. Slide one side into the other, overlapping layers on the top and the bottom.

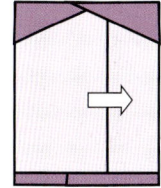

9. Inside view of the sliding. Both on top and bottom, flaps fit into pockets similar to the more tubular aircraft.

10. Roll the bottom up into the base to secure the pylon.

11. Side view. Reverse fold the tip back.

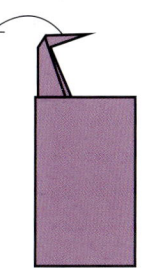

12. And reverse fold it back the other way.

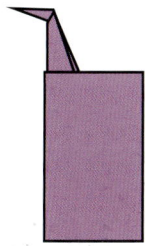

The pylon stand is now ready to display airplanes.

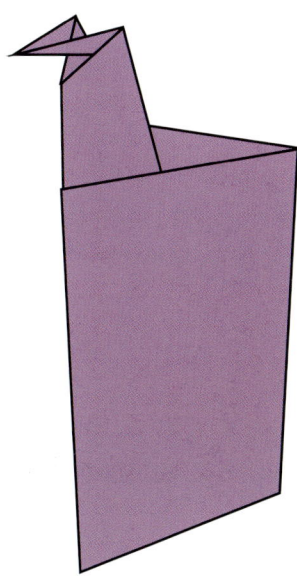

Flying is hypnotic, and all pilots are willing victims to the spell. — Ernest K. Gann

Stands

Hand Stand

Inspired by the madness of my good friend Jeremy Shaefer. Begin with a square much larger than the airplane it is holding, colored-side up and creased down the middle.

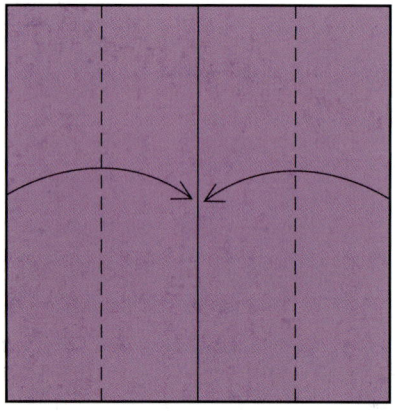

1. Valley fold both outside edges into the center. In origami jargon this is called the cupboard fold. You don't want to hear about the toilet fold.

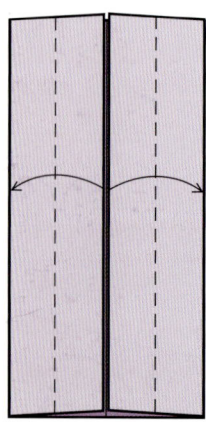

2. Valley fold both raw edges to meet the outside folded edges. In origami jargon this is called the damaged cupboard fold.

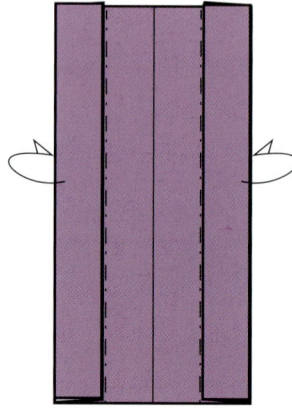

3. Mountain fold so that the outside folded edges lie on the center.

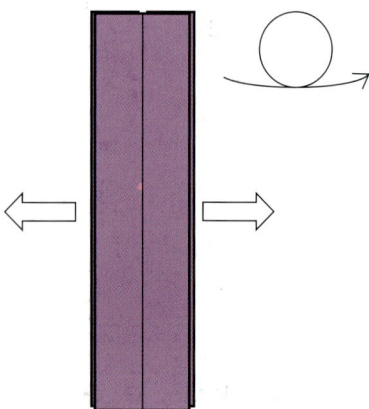

4. Unfold to step 1 and turn over.

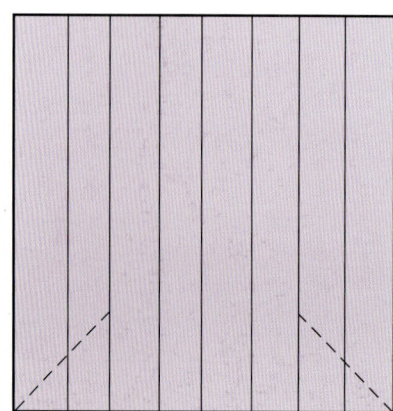

5. Crease the diagonals partway.

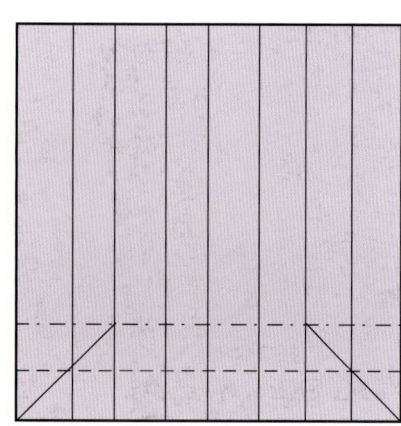

6. Make the pre-creases shown.

120

Hand Stand

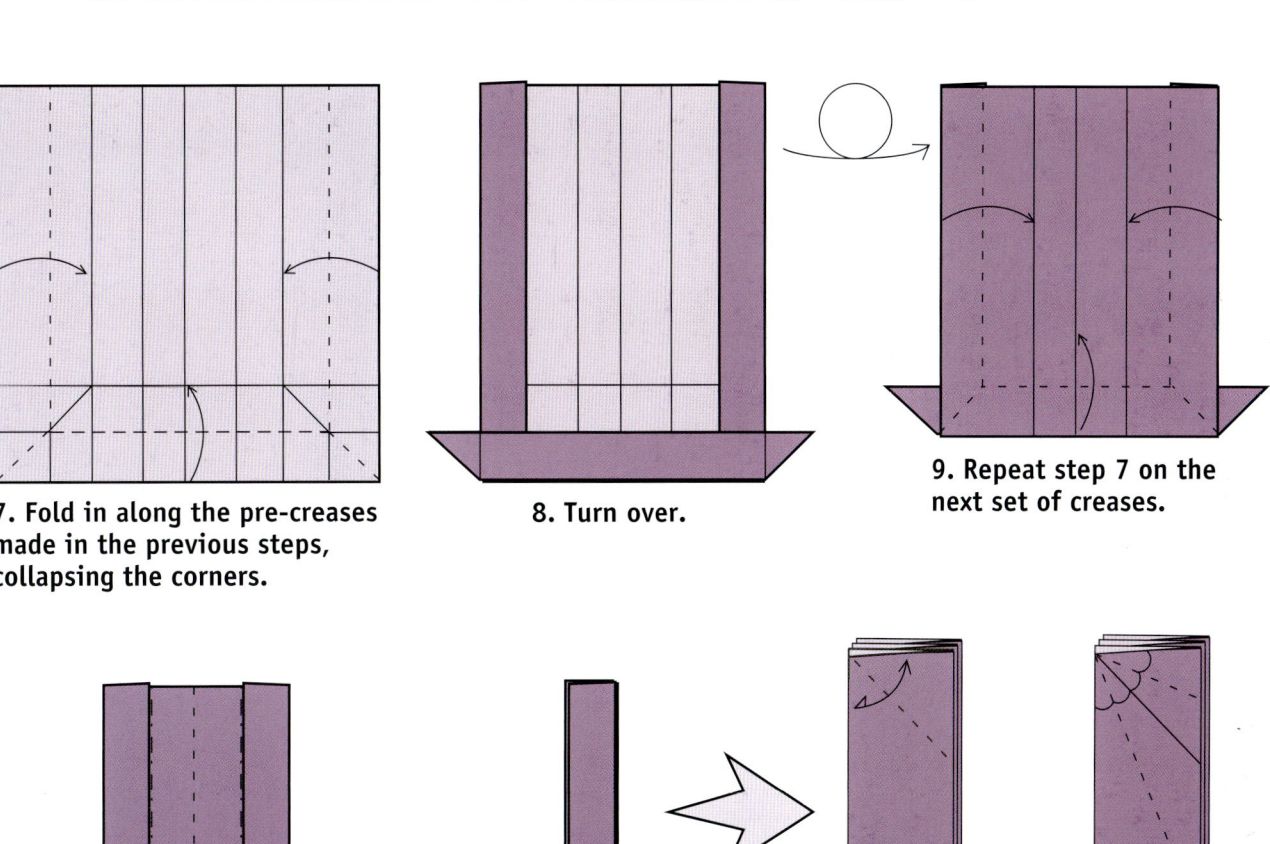

7. Fold in along the pre-creases made in the previous steps, collapsing the corners.

8. Turn over.

9. Repeat step 7 on the next set of creases.

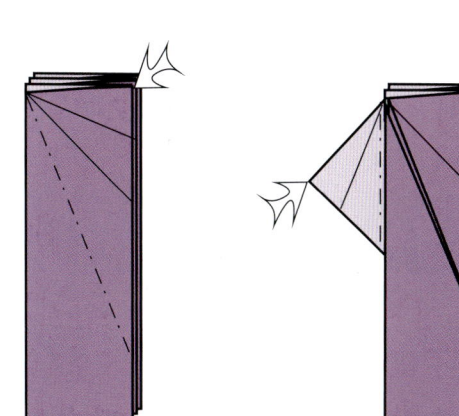

10. Fold back and forth along preexisting creases.

11. The top will be shown enlarged.

12. Crease the diagonal.

13. Crease the angle bisectors.

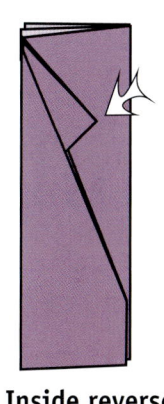

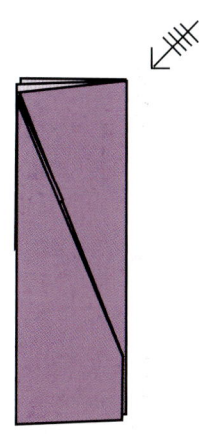

14. Inside reverse fold along the first crease.

15. Inside reverse fold along the next crease.

16. Inside reverse fold one more time.

17. Repeat steps 12–16 on the three remaining flaps. Please don't hate me.

121

Stands

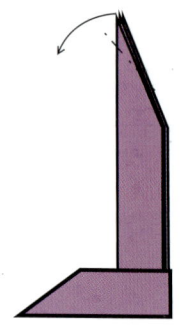

18. Outside reverse fold the ends for fingers.

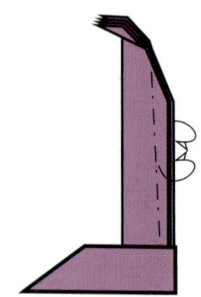

19. Thin the hand by folding the back flaps into the stand.

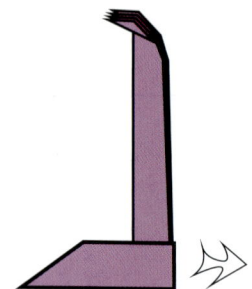

20. Pull out paper from the back to complete the stand. You will need to squash fold some paper on the inside.

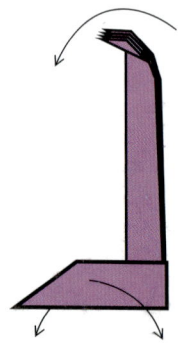

21. Spread the fingers a bit, and fold out the anterior flaps (you do want your stand to stand, don't you?) to complete the Hand Stand.

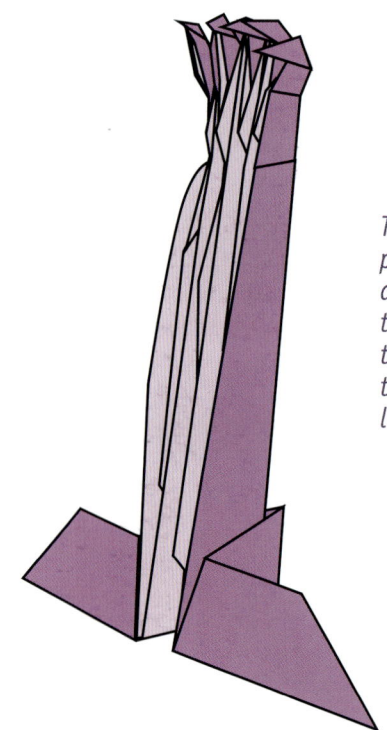

The fingers of the Hand Stand can be put into some of the pockets of the aircraft. It's especially good for the tubular models. Remember to make the Hand Stand out of larger paper than the airplane. The hand is far larger than the airplane it's throwing.

When once you have tasted flight, you will forever walk the earth with your eyes turned skyward, for there you have been, and there you will always long to return. —Leonardo da Vinci